STAY COOL

STAY COOL

Why Dark Comedy Matters in the Fight Against Climate Change

AARON SACHS

NEW YORK UNIVERSITY PRESS

New York, New York

NEW YORK UNIVERSITY PRESS
New York
www.nyupress.org

Please contact the Library of Congress for Cataloging-in-Publication data.
ISBN: 9781479819393 (hardback)
ISBN: 9781479819409 (library ebook)
ISBN: 9781479819423 (consumer ebook)

New York University Press books are printed on acid-free paper, and their binding materials are chosen for strength and durability. We strive to use environmentally responsible suppliers and materials to the greatest extent possible in publishing our books.

Manufactured in the United States of America

10 9 8 7 6 5 4 3 2 1

Also available as an ebook

For Sam, Abe, and Ozzie

If heat rises, then heaven might actually be hotter than hell.

—Steven Wright

CONTENTS

PROLOGUE

Paradise Not Included

The darkest comedy is often about hope or faith. But the goal isn't any kind of triumphal transcendence—it's more like endurance and persistence.

In the summer of 2012, the comedian Tig Notaro started a stand-up routine with the line, "Good evening. Hello. I have cancer, how are you?"[1] And she really did have cancer: the diagnosis had been delivered about a week earlier. Right after her mother had died.

Must the show go on?[2] Notaro was shell-shocked by her diagnosis; at first she just felt like hiding. But after a couple of days, she decided that she wanted to take the stage one last time before having surgery. She wanted to feel like herself, wanted to persevere, wanted to see if she and her audience could share some laughs about the worst things that had ever happened to her. Maybe the laughter would even help her cope. She sensed that the public performance of her grief and fear could shift the stories she was telling herself in private. And she hoped it would be liberating for the audience as well.

It worked. She felt that her life had been pretty great until recently—so she was thinking about her upcoming mastectomy as just "tit for tat." She joked about how the hospital had sent her dead mother a questionnaire asking how her stay had been (not spectacular). "During this hospital stay, did nurses explain things in a way you could understand . . . considering you had zero brain activity?" She joked about how she had also just been through a breakup, so now she was trying to create a new online dating profile: "I have cancer. Serious inquiries only." And, after a couple of awkward minutes, the audience was right there with her. As soon as they registered her bad news, she was able to accept it in a new way, and she could start trying to make *them* feel better instead of dwelling on her own anxiety and anguish. They were all in it together. "It's OK," she said to them. "It's gonna be OK. . . . It might *not* be OK. But I'm just saying. It's OK." The surgery was successful; a recording of the show went viral; her career took off; she still worries about when the cancer might come back.[3]

Worrying is part and parcel of modern life—especially now, in the age of climate change. We don't all have cancer (yet). But we all have global warming. And almost everyone who thinks about it finds it utterly overwhelming.

For years, those of us trying to fight climate change focused on the obstacle of denialism, and there are still a few key politicians who need convincing (mostly white men of a certain age in a certain party).[4]

But as more and more people have accepted the reality of the problem, the central challenge has shifted. Now what we really have to deal with is our immobilizing despair and desperation. Mention "climate change," and many people immediately conjure up heartbreaking, terrifying images of all-consuming fires, sweltering cities, crashing floodwaters, desiccated fields, and ragged crowds of refugees suffering from heat exhaustion. How are we supposed to cope with all *that*?

I'm just saying: dark comedy could help.[5]

Perhaps not in a perfectly straightforward way, though. Comedy is one of our strangest, most complicated, and least understood inventions. Sometimes you're in the mood for it and sometimes not. I can attest that the same joke can be sidesplitting to a Jew from Boston (like me) and contemptible to an ex-Mormon from Los Angeles (like my wife). Comedy is an unstable genre: How else could it encompass Shakespeare, Richard Pryor, Tina Fey, the Three Stooges, and puns?[6]

For some people, comedy just isn't appropriate for certain topics: it's always too soon. The effect will always be too unpredictable. Tig Notaro was terrified before she went onstage to do her cancer set. Her audience had come to laugh, not cry, and the news she was bringing was definitely not inherently funny. But the edgiest comedians know that the distance between laughing and crying can be quite short and that comedy often goes better when it confronts difficult truths. In the early months of the COVID-19 pandemic, even with infection rates spiking, performers gradually carved out space to note the ridiculousness of various lockdown experiences. One *Saturday Night Live* sketch acknowledged the enhanced stress of family life by proposing that parents should let their kids drink, too. (My wife and I were a couple of weeks ahead of the *SNL* writers.) As nice as it is to escape into frivolity for a while, most of us would also welcome the chance to gain some purchase on reality. The combination of comedy and tragedy, as Notaro suspected, can sometimes produce a new kind of community, a new kind of catharsis.

And as Notaro well knew, there is a long tradition of dark comedy (sometimes called "black comedy"—not to be confused with African American comedy)—which people have clung to even in the most horrific circumstances.[7] Numerous Holocaust survivors testified that a sense of humor was crucial for morale in the concentration camps: "You shouldn't eat too much," a group of friends used to say to each other in Treblinka, "because we're the ones who will have to carry away your body."[8] Laughing is a survival instinct, even if it doesn't always feel instinctual.

As a historian, I was at least dimly aware of the dark-comedy tradition, but until recently, that mode simply hadn't occurred to me as a potentially useful tool in the climate struggle. I was busy with the ultraserious problem of Americans' inability to acknowledge that global warming is happening and that we have an ethical responsibility to counter it. In particular, I was struck by the parallel between our denialism surrounding environmental issues and our denialism surrounding death: Americans are terrible at recognizing limits. But my historical research suggested that this hadn't always been the case. Before the catastrophe of the Civil War, US culture was obsessed with the limits imposed by nature, and death was woven into both everyday life and everyday environments, in incredibly generative ways. So in 2013 I published a book that I thought of as an inoculation of sorts: by questioning our current culture of denial and celebrating an older culture that had developed better strategies for facing mortality, I hoped to nurture more resilience in myself and others, for all of our inevitable confrontations with limits, from the death of loved ones to climate tipping points.[9]

Just four months after my book came out, my father died, from complications related to Alzheimer's. And then my mother, who had been caring for him for years, sank into a dementia of her own.

At first, for about a year, I was in a good position to handle my sense of loss and also to support my mother, because a research fellowship had allowed me to move to Boston, where she was still living, with in-home assistance, in the house where my sister and I had grown up. But then everything fell apart.

My mother went from being quiet and gentle to being violent and abusive. It had felt hard enough to witness my father's decline: a retired professor, he had been not only wise and thoughtful but meticulously considerate, until his dementia reduced him to a wreck of confusion. My mother's situation, with my father already gone, sent me into a spiral of depression. I was exhausted, grief stricken, and constantly frustrated that it was so difficult to figure out how to help. Our society is not well equipped to deal with the spiritual and moral challenges of aging. Talk about denial.

When my mother died, in September 2015, I felt a measure of relief—and then I felt bereft. Between the summer of 2014 and the summer of 2017, I was living in a Purgatorial haze of guilt, sorrow, and back pain. For most of my life, I had coped with tension and angst by taking long walks and reading long books—both activities that reminded me of the wider world and greater purposes, of intellectual and environmental diversity, of the possibilities of creation. Neither strategy worked anymore. It seemed completely out of the question to pick up the writing project I had launched during my year in Boston, which was about—wait for it—"the trauma of modernity." I was no longer impressed by the writers who figured out almost two hundred years ago that industrial capitalism's fetishization of stability and security was ultimately going to make most people feel more unstable and insecure. You could not have paid me to read another Melville novel.

Meanwhile: Hurricanes. Floods. Wildfires. Refugees. Trump. It felt like the End of Days. And everyone was either depressed or in a panic. No one was stepping back to take a longer view. No one was laughing. It seemed absolutely impossible even to think about climate change without sliding into an emotional abyss. The perpetually gloomy students in my environmental history courses seemed desperate for any morsels of good news, which I was hard-pressed to cook up.

Those three years of my life were what Melville would call my Cape Horn. (Yes, I'm reading him again.) "But, sailor or landsman," he wrote, in the novel *White-Jacket*, "there is some sort of a Cape Horn for all. Boys! beware of it; prepare for it in time. Graybeards! thank God it is passed. And ye lucky livers, to whom, by some rare fatality, your Cape Horns are placid as Lake Lemans, flatter not yourselves that good luck is judgment and discretion; for all the yolk in your eggs, you might have foundered and gone down, had the Spirit of the Cape said the word."[10]

During that time, one thing besides family, friends, and anti-inflammatories kept me afloat: dark comedy. I used to recline in my portable zero-gravity chair and laugh and cry at the same time while I watched hour-long stand-up specials. I developed a deep gratitude for the comedic traditions that have persisted in our culture by making light of our most painful and stubborn problems.

In February 2015, my doctoral advisee Amy Kohout invited our mutual friend Jenny Price to give a talk on our campus, based on her book project, called *Stop Saving the Planet!*[11] And it was funny. And Jenny told me about a course she'd been teaching on environmental humor. She was using comedy to help people recognize that it might come across as kind of insensitive to focus on protecting the Planet and Its Future when there were lots of vulnerable human beings suffering from the effects of climate change right here and now.

Jenny and I agreed to collaborate on a sketch-comedy panel for the next environmental history conference, which wound up including Nicole Seymour, whose 2018 book, *Bad Environmentalism*, is full of excellent examples of irreverent, charged humor.[12] Ultimately, Jenny and Nicole showed me how strategically important it could be for environmentalists to lighten up. The personal and the political had come together in the most powerful way I could have imagined.[13]

I started thinking more deeply about the role of morale in social movements, about how peasants and slaves and the homeless and the oppressed and activists and artists have miraculously turned despondency into hope, determination, and defiance. Sometimes it's religion that keeps people going; sometimes it's music; sometimes it's comedy.[14] "Indians have found a humorous side of nearly every problem," wrote Vine Deloria Jr. (Standing Rock Sioux), in his book *Custer Died for Your Sins*, which helped to galvanize the American Indian Movement in the late 1960s and early 1970s. "The more desperate the problem, the more humor is directed to describe it. . . . When a people can laugh at themselves and laugh at others and hold all aspects of life together without letting anybody drive them to extremes, then it seems to me that that people can survive."[15] Dark humor can be an incredible spur to resilience. No amount of progress on any front will negate the need for good, humane communities to stand together and support each other in the effort to laugh—in the face of evil, disaster, oppression, structural racism, death, taxes, and department meetings.

A year after Jenny's visit, when a colleague asked me to step in for someone and give a lecture on a subject of my choosing, I decided to gather some environmental comedy material and try it out. For the sake of the talk, I pretended that I was working on a new book project: it would be based on Dante's *Divine Comedy*, but my book would explore the humor of environmental decline—you could call it "Decline Comedy." Several students came up to me afterward and asked if I would give a similar talk to other groups on campus that they were a part of. Several also asked me when my new book was coming out. They said they felt relieved to laugh at a problem that had always just depressed and overwhelmed them; they felt like they had a new language with which to talk about climate change.

I was stunned. My words had never before seemed so activating.

I wound up delivering a few different versions of that lecture, which I started calling "The Climate Change Comedy Hour," and the response was phenomenal. People kept telling me that I had cheered them up, given them a different kind of hope. And eventually that cheered me up. And then I started writing this book, which I hope will cheer you up.

That's my main goal. Dark comedy can be useful in a number of different ways, but in the specific context of this book, its most crucial contribution is as a morale boost, for anyone confronting the darkness of climate change. Activists can also use self-denigrating dark comedy to make themselves more relatable and satirical dark comedy to attack their enemies, as I hope will become clear. But at the core of this book is my conviction that simply knowing some relevant, funny history could help you cope with your despair—which might be the first step toward creating a broad, inclusive climate movement.

At the same time, of course, I should emphasize that this is not the Feel-Good Story of the Millennium. It's more like a guide to reframing Purgatory.

Purgatory is not a bad option. History suggests that Purgatory is basically what humans have been trying to get used to since time immemorial. Isn't it at least better than the apocalypse? Just before the final tidal wave (spoiler alert) in the film *Don't Look Up* (2021)—perhaps the most prominent example of climate comedy to date—the lead character says, "We really did have everything, didn't we?" Well, everything except the political will to reduce carbon emissions.[16]

After waiting sometime in purgatory, Beryl decides to redecorate.

Life sometimes feels pretty awful for a lot of people in the twenty-first century, but things could definitely be worse, and they often have been. Of course, I can understand why Purgatory would sound disappointing, especially if you're still dreaming of Paradise. But Paradise is a boondoggle. Why not just dispense with it? The Garden of Eden was invented to make us feel guilty. We've always had to contend with serpents, not to mention heatstroke and sunburns.

The structure Dante chose for the *Divine Comedy* was perfectly reasonable at the time. For a medieval Italian Catholic, it made sense to start with a tour of Hell, then move through Purgatory, and finally give readers something resembling a happy ending by bringing them to Paradise, where they might catch a glimpse of God. Back then, "comedy" referred less to humor and more to a faith that Heaven and Earth were aligned and justice would be served in the afterlife.

But this is *decline* comedy. As a Reform Jew and an environmental historian, writing seven hundred years after Dante, I say we embrace Purgatory.

I've always thought that the human condition was best captured by the character of Sisyphus, perpetually rolling his boulder up the hill and then watching it roll down again just before he reaches the summit. And I've always agreed with the argument made by Albert Camus that Sisyphus might actually be happy: he gets lots of exercise; he works outside; and, most of the time, he feels as though he's getting somewhere. It's better than getting crushed by the boulder, right? Yes, Camus was killed in a car crash, but that doesn't negate his philosophy. And yes, I also know that Sisyphus is supposed to be pushing his boulder in Hell, but I'm choosing to imagine him in Purgatory, catching a glimpse of the sunset.[17]

Climate change is just the latest Divine Punishment or Twist of Fate or Proof of Human Stupidity, depending on your perspective. As Karl Marx once said, approximately, "History repeats itself, the first time as tragedy and all the other times as farce."[18] Even though climate change feels new and different, we've seen this kind of crap before. We defeated the Black Death; we abolished slavery; we survived disco.

History has certainly brought us its fair share of floods, plagues, and wars, but it has also taught us how to laugh in the face of such calamities. And how to move on. And rebuild. A few good doses of history and comedy can often put things in perspective. The people of Hiroshima regularly commemorate the devastation of their city with a song reminding them that "You must tell a story of your hard times / And laugh twice."[19]

Of course, history and comedy are not panaceas, and they won't make everyone feel the same way. Comedy can seem just as fatalistic as tragedy. You laugh for a second—and then what? What if you don't feel spurred to action? Or what if comedy actually makes things worse, because we wind up merely laughing *instead of* setting up barricades in the street or, even better, building flood walls?[20]

The comedian Hannah Gadsby had critics tying themselves in knots in 2018 when her "uproariously funny" stand-up special, *Nanette*, made an explicit argument against . . . comedy. After gently poking fun at her own seeming "masculinity" for a few minutes, she suddenly started expressing the fear that those sorts of jokes might allow people to take misogyny and homophobia too lightly. She even went so far as to say that, in order not to make things worse, she was probably going to stop being a comedian. But that message, too, was delivered comedically, so that, in the end, what she really seemed to be saying was that maybe the best comedy is the kind that attempts to be funny and serious simultaneously. To which I say, "Amen," and to which the critics said, "Wait, does that mean we should call it 'Anti-Comedy'?"[21]

In the realm of climate change, we clearly haven't gotten very far by invoking fear and tragedy. So why not try something different? Why not spend some time with a few tricksters—with satirists, with clowns and jesters, or even with fools? Comedy revels in surprises, in new perspectives, in making the world topsy-turvy, carnivalesque. Imagine Al Gore in a hall of mirrors.[22]

Credit where credit is due: Al Gore *can*, on occasion, be funny. He started out as the most wooden of politicians and the most self-righteous of environmentalists. But eventually he learned how to give a reasonably witty slide show (at one point in *An Inconvenient Truth*, he mentions that "denial is not just a river in Egypt"). He also figured out that he could help his own cause by making fun of himself. Have you ever seen him do the Macarena? You could look it up.[23] More recently, he promoted the sequel to *An Inconvenient Truth* as a "hot date movie" (emphasis on "hot"), and he teamed up with Stephen Colbert to offer "Al Gore's Climate Change Pick-Up Lines," all of which played on his own earnest wonkishness: "Looking at you, two things are clear: heaven is missing an angel, and the US is missing any kind of viable, responsible climate policy."[24]

Humor is so crucial to the climate movement precisely because we environmentalists have so rarely made fun of ourselves. That's the argument of chapter 1 of this book: it works its way through the dour and sanctimonious history of US environmentalism, critiquing it along the way, and then juxtaposes that narrative of righteous doomsaying with the more self-critical history of US comedy. Self-mockery is one of the most useful and yet most underrated forms of dark humor. But in the mid-twentieth century, comedians in the

Al Gore, looking serious

US figured out that often the key to landing an edgy, political point was simply coming across as more human. "I make my living by self-denigration," Phyllis Diller once said; "then I go on to the denigration of others."[25] Environmentalists somehow never recognized that self-undercutting humor is a great technique to use when you need to speak compellingly about overwhelming problems. A 2013 study published in a journal of social psychology confirmed that a lot of environmental activism fails simply because people "have negative stereotypes of activists"—stereotypes that, alas, are well earned.[26] Already, some of the best climate comedy has made its mark by playing on environmentalists' tendency to sound as if they hate all human beings (including themselves), as in *The Onion*'s advertisement for a new Toyota Prius model that automatically impales its owner on a "killing spike" in order to reduce their carbon footprint to zero. (You can find it online: the Prius Solution.) "If you care about the planet *at all*," says one citizen who has preordered the vehicle, "it's the best car you can get. If your car doesn't kill you, it's like what's even the point?"[27] Making fun of ourselves could also help us remember what the point really is: not saving the planet but rather coming together to help the planet's most vulnerable people and the environments they depend on.

Chapter 2, the Purgatorial heart of the book, turns toward the long comedy traditions developed by African Americans and Jews, people who have often dealt with awful situations and who teach us that just because you're languishing on the edge of Hell doesn't mean you can't put up a fight. Indeed, gallows humor, which quite often merges with satire, is the best-known form of dark comedy and a well-established means of bolstering morale.[28] No, the enslaved people who made fun of their masters did not bring down slavery, but they did help sustain resistance and hope—just as civil rights activists did when they traded jokes about powerful white supremacists like Governor George Wallace of Alabama.[29] I learned this lesson most directly as a child by watching the 1971 film version of *Fiddler on the Roof* over and over again—you know, that joyous musical in which Jewish weddings get interrupted by pogroms and ultimately the entire Jewish community gets expelled from their Russian shtetl. The Jews respond to the edict of expulsion by singing the song "Anatevka," whose basic message is: we never liked living in this piece-of-crap village anyway, with its tumbledown shacks and oafish Russian soldiers. The song is slow and sweet, a eulogy and an elegy both, and yet it also, miraculously, manages to be defiant, by undercutting its tragic theme with comedic asides. "Well," says one character, "Anatevka hasn't exactly been the Garden of Eden." What do you do when you're suddenly forced to pull up stakes? When you have to improvise? The tragic mode is not particularly good at generating improvisation. But comedy is. Those Jews were going to move on and survive. And couldn't that attitude be immediately useful in today's climate crisis? For example: would it really be so terrible if we had to abandon Florida to its climate-change-denying Republicans?

There's been a spike in "flooded Florida" jokes recently, and chapter 3 argues that in fact climate comedy, like sea levels, is on the rise. Comedians, and even some activists, have come up with many novel combinations of self-directed comedy, gallows humor, and dark satire, in their efforts to address global warming. And it's not just about adapting civil rights humor and shtetl humor. Even though environmentalism has often been earnest and grim, there is actually a long history of dark comedy explicitly about nature that we can draw from. After all, nature has a pretty twisted sense of humor: it provides the basic conditions of our survival, but it won't hesitate to send a spur-of-the-moment tornado toward you while you're walking down the street with your hands in your pockets, whistling a tune. So certain people, when they weren't too busy being oppressed and exploited by other human beings, made time to joke about the violence inflicted by nature, even during famines and epidemics and other environmental emergencies. The polar explorer Sir Ernest Shackleton often asked his prospective crewmates if they liked to sing, because he knew from experience that the occasional musical revue or drag show could go a long way in sustaining morale if your ship were to get stuck in the ice.[30] Alas, polar exploration just isn't the same in the age of Global Melting. But polar-bear jokes are circulating as liberally as Arctic waters, and that's probably cause for hope.

A number of recent articles and books have demonstrated the value of all kinds of humor in activating people—especially millennials and Gen Zers, who grew up getting their news from "reporters" and "pundits" like Stephen Colbert. One 2018 study had the title "Not Just Funny After All: Sarcasm as a Catalyst for Public Engagement with Climate Change." (Tapping into a sense of outrage can also be a good activation strategy, though only if your audience isn't already too despairing or anxious: taking a pissed-off and determined tone won't help you reach people who are hiding in the storm cellar while contemplating a move to Canada.)[31] In any case, comedy almost always works better than scaring people to death or scolding them, especially when the people you're talking to have already endured plenty of scaring and scolding. It can be fun for everyone when you rework old narratives and perform them in new ways. So even though I'm a huge fan of Greta Thunberg and her climate strikes, I'm hoping that her followers will gradually move away from her Bergmanesque affect and maybe go more in the direction of the *Muppet Show*'s Swedish Chef. The physiological and emotional benefits of laughter are well documented, and people who laugh together might well be more likely to attend a climate-justice rally together.[32]

Of course, there is no rock-hard evidence that certain kinds of jokes will always inspire certain kinds of political behavior.[33] The communication studies scholars studying climate messaging have tended to uphold pleasant, "good-natured" humor as the best way to reach people, and they may well be right about many of our fellow citizens.[34] But that kind of comedy might come across to others of us as too safe, or even naïve, under current circumstances. Now that most of us recognize the truth of climate change, we're going to need our comedy darker, more bitter, so that we feel fully acknowledged in our awareness of the dark, bitter realities that we're facing. Dark comedy has not been as thoroughly studied or theorized as other forms of comedy, so it may seem like something of a loose cannon—but, to me, this feels like a good moment to explore its possibilities.

It is undeniable that, at the very least, dark comedy can open up space in culture, and more capacious cultures are better able to handle the kinds of issues that people often turn away from. In my lifetime, things have clearly improved for the LGBTQ community (though of course there is still much work to be done on that front); such progress reflects both the direct labor of rights activists and also the subtler labor of people like cartoonists and TV producers. We're living in an era when charged, potentially activating humor is all around us, not just on late-night television but even in prime time. You might or might not find shows like *Will and Grace* or *Modern Family* particularly funny, but if you've spent time watching them, you've at least been forced to consider the implicit argument that gay couples and straight couples have a lot in common: one partner will always do less housework (guilty as charged), and one will always take forever to leave parties (I'm looking at you, Honey).

When it comes to climate change, many people feel simultaneously like victims and perpetrators: we are all affected, and we are all complicit—in different proportions, yes, but perhaps the commonality is what's most important. As some leaders emphasized at the beginning of the COVID-19 pandemic, times of danger should help us remember that we're stronger together. We need to reach across every form of difference: only a less traumatized, less divided citizenry will be able to replace the fossil-fuel economy.[35] Our long traditions of dark comedy suggest the need to confront awful realities with flexible attitudes and mind-sets—with a willingness to laugh at ourselves even amid immediate threats. The

Youth International Party, or Yippies, led by the Jewish activists Abbie Hoffman and Jerry Rubin in the late 1960s and early '70s, could not claim many political victories, but their surrealistic pranks helped remind people that one way of countering the rigidity of racism and militarism and industrialism is simply to remain supple. Running a third-party pig for president in 1968, the Yippies earned a great deal of goodwill by capturing the resentment that both Republicans and Democrats felt toward the unbending political establishment.[36]

Dark comedy has actually been a source of solidarity for millennia—as I realized recently when I found a prime example in the Bible. I don't think I've ever quoted chapter and verse before (like I said, I'm a Reform Jew, which is essentially another term for Unitarian), but it's right there in Exodus 14:11–12.[37] When the Jews successfully escaped Egypt, only to be halted by the Red Sea, they felt sure that they would die right there: Pharaoh's army was bearing down on them. But their response was to shake their fists at the heavens and ask God, "What was the point of helping us flee, if we were just going to die here in the wilderness? Were there not enough grave sites in Egypt?" Remarkably, the Lord appreciated the joke and proceeded to part the waters. Which just goes to show: if you get really good at gallows humor, you might even be able to control sea levels.

1

INFERNO

Environmentalists, Feel the Heat!

(Or, The Usefulness of Self-Directed Humor, Especially When You've Been Pontificating about Looming Disasters for More than a Century)

The good news: the planet is going to be just fine. It's been hot before. Fifty million years ago, North America was covered in jungles and swamps, which turned out to be excellent habitat for anteaters and carnivorous lizards.[1] Life will go on, in some form or another.

It's just the human beings who seem to be going to Hell. Some sooner than others, of course. If you live on a mountain and have a relatively reliable solar-powered air-conditioning system, you should be fine for a few more decades. So go ahead and reproduce.

Then again, if you live off the grid on top of a mountain, you're probably either retired and have already done your reproducing, or you're a hipster who was never going to reproduce in the first place, since kids are definitely not environmentally correct.

> Q: How do you know when you're in a room with environmentalists?
>
> A: Oh, they'll let you know.

If you're the kind of person who would pick up this book, then you're probably something of an environmentalist yourself and also someone who has noticed how ridiculous we environmentalists can be. We can be pretty ridiculous.

On the television show *Modern Family*, Mitchell, an affable, neurotic, gay environmental lawyer, is one day confronted by a neighbor (played by a relatively famous indie actor). "I think there might be something wrong with your air conditioner," the neighbor says, "because it seems to be running a lot even though it's kinda cool outside." Mitchell tries to laugh him off: "Oh, it's not broken—it's just that my partner runs a little hot." The neighbor doesn't miss a beat: "Not as hot as our planet."

And then the show's writers gradually turn up the heat on this little confrontation. The neighbor, seeming to show some self-awareness, apologizes: "I don't mean to be That Guy." But of course that's precisely who he means to be. Mitchell, still trying to be nice and also hoping to save face, notes that at least he drives a Prius. The neighbor one-ups him by explaining that *his* car runs on reclaimed vegetable oil—which finally triggers the prosecutor in Mitchell, who launches his own offensive about home energy use. But the neighbor reveals that he's gone 100 percent solar: he actually sells electricity back to the grid.

The writers' sympathy—and therefore the audience's sympathy—is clearly with Mitchell, but the neighbor gets in some great jabs and gibes. When Mitchell tries to shrug off his flaws and claim that he's still "pretty green," the neighbor looks down, pointedly, and says, "So is your lawn." I don't know if there's a perfect word to describe Mitchell's expression—equal parts embarrassment and exasperation—but it's recognizable to me as the look you might get on your face when you realize that your parents were right all along about one of your key shortcomings, which they are now proceeding to lecture you on at length, in front of your own children. In this case, the neighbor also happens to have the perfect yard-care solution for Southern California: "I went drought-tolerant: succulents, indigenous plants, rock garden." And then, suddenly, a miracle happens, and Mitchell's daughter, who has in fact witnessed this whole exchange, comes to his rescue, by channeling Mitchell's much-less-diplomatic husband and telling the neighbor where he can stick his cacti. "My other daddy," she notes, matter-of-factly, "says your yard looks like a litter box." Ah, the sweet revenge of ordinary people who want to be as environmentally friendly as possible but who still retain some sense of aesthetics and a desire to create spaces where children might actually want to play.

The scene should really end there. But the writers decided to give the last laugh line to the neighbor: "Cute kid," he says, totally unflustered. "I remember when she was in disposable diapers."[2]

Mitchell looks at the camera for a second, and we know exactly what he's feeling: that static combination of shame and annoyance. He's stuck. As are we: we know we're not doing enough, but there's no way we're going to become That Guy, because That Guy is a jerk. Why are environmentalists so damn righteous? Why don't we have any decent role models?

> Q: Why do Prius drivers keep only one hand on the steering wheel?
> A: So they have the other hand free to pat themselves on the back.

If you want to support people of color and civil rights, you can go to a Black Lives Matter march. If you want to protest the income gap, you can Occupy Wall Street. If you're a feminist, you can join the #MeToo movement (or, if you're a guilt-ridden man, the #IHearYou and #HeForShe and #AllMenCan movements). If you're painfully aware of the privilege of heteronormativity, you can participate in a rally promoting equality for gay people like Mitchell. Why do climate marches feel so much more complicated?

Back in 1993, I went to a gay-rights demonstration in Washington, DC, and within minutes, I was marching with a throng of like-minded people and joining in their rhythmic chant: "We're here! We're queer! We're here from Tennessee!" The only thing about the chant that was true for me was that I was there. But so what? It was all about presence and solidarity, and anyway, we're all a little bit queer, and we all, at some point, have to reckon with the fact of Tennessee. "We're here!" I yelled, with pride, with conviction, with a firm belief in the absolute undeniability of what I was saying.

That kind of commonality amid difference is much harder to achieve at a gathering of environmentalists. Our "movement" is not only internally divisive but incredibly, well, immobilizing. It might not matter where you're from, once you arrive at a climate march—but it matters how far you traveled and what form of transportation you used. You damn well better have walked or biked, or you're going to face some hard questions. What's the carbon footprint of your skateboard? For the planet's sake, I hope that bus was powered by reclaimed vegetable oil!

Most of us are so paranoid about being judged for our indulgent lifestyle choices that we just decide not to attend the marches at all. What's the point? It would be more environmentally friendly to stare at the cacti in your rock garden.

We deserve to be mocked. Feel the heat, environmentalists!

I had moved to Washington, DC, after college to take a job with an environmental organization, the Worldwatch Institute, where everyone in the office shared the same web password: "DOOM."[3] (And only one person could use the internet at a time: it was

1993.) It's now three decades later, and the internet has changed everything—except environmentalist discourse.

Consider this list of recent book titles:

Learning to Die in the Anthropocene: Reflections on the End of a Civilization
Down to the Wire: Confronting Climate Collapse
The Collapse of Western Civilization
The Uninhabitable Earth: Life after Warming
Last Hours of Humanity: Warming the World to Extinction
Requiem for a Species: Why We Resist the Truth about Climate Change
Storms of My Grandchildren: The Truth about the Coming Climate Catastrophe and Our Last Chance to Save Humanity
High Tide on Main Street: Rising Sea Level and the Coming Coastal Crisis
All Hell Breaking Loose: The Pentagon's Perspective on Climate Change
Climate: Code Red

I think of the authors of these books as my colleagues, and I admire and appreciate each of them. But they are relentlessly unfunny.[4] As were virtually all of their predecessors in the environmental movement.

Take John Muir, founder of the Sierra Club and First Propagandist for National Parks. Great activist, but he was the opposite of funny. Aldo Leopold, author of *A Sand County Almanac* and First Prophet of Ecological Restoration? Inspiringly wise but dead earnest. Rachel Carson? One of my favorite writers ever; but try to find a single humorous line in *Silent Spring*.

It's true, though little known, that Henry David Thoreau, all the way back in the mid-nineteenth century, was actually quite hilarious. Reviewers of his writings and lectures spoke of his "delicate satire against the follies of the times" and noted that he was often able to "keep the audience in almost constant mirth," that he even succeeded repeatedly in "bringing down the house by his quaint remarks."[5] Once, in a letter, he asked the rhetorical question, "What is the use of a house if you haven't got a tolerable planet to put it on?"[6] Thoreau seemed to recognize the human condition as fundamentally Purgatorial. What right do we have to expect anything beyond tolerability?

On the subject of well-intentioned activists, Thoreau said, "If I knew for a certainty that a man was coming to my house with the conscious design of doing me good, I should run for my life."[7] Most people think that Thoreau ran from pretty much everyone, but in fact he was quite social. It took a special level of pompous hectoring to trigger his flight response (though it must be acknowledged that he was perpetually annoyed by his hypocritical neighbors who claimed to be opposed to slavery but never spoke out against it). Generally, Thoreau was willing to hang out with anyone else who acknowledged the human tendency toward narrow-mindedness, destructive selfishness, and general bungling. "Thank God, men cannot as yet fly," he once said, as if to a friend on the porch, "and lay waste the sky as well as the earth!"[8]

Note to Thoreau: we can fly now. Not that he would care: "Probably I should not consciously and deliberately forsake my particular calling to do the good which society demands of me, to save the universe from annihilation."[9]

Alas, Thoreau's dark wit never really caught on among his fans. As Americans developed something like an environmental movement at the turn of the twentieth century, they embraced an anxious, alarmist, accusatory mode that, at the time, turned attention away from their own complicity in the crisis they were decrying. You can reduce their rhetoric to two exclamations: "We're running out of the natural resources that made us rich!" and "Soon we might not have any more beautiful places where we can go hiking and birdwatching! Especially if people keep killing birds and putting them on hats!"[10]

Turn-of-the-century bird hat

Ah, privilege. There were of course some admirable aspects of early environmentalism: the ecological emphasis on interdependence and community, the concern for public health, the national parks' excellent signage. But let's face it: when your main goal is to protect the environment for your own use, without taking any responsibility for its endangerment, then your tone is probably going to stay pretty serious, because you need to work so hard to win the moral high ground (insert Sierra Club mountaineering joke).[11]

In other words, environmentalist seriousness is not a new phenomenon; it's not tied to the seriousness of climate change. The sky has been falling for well more than a century.

"As the destruction of forests goes on," wrote the reformer Joseph Rodes Buchanan in 1890, in an article titled "The Coming Cataclysm of America and Europe," "our floods increase in power, and large regions are threatened with barrenness." Ten years later, another tree-hugger used similarly threatening terms to lecture his fellow Californians: "The deserts even now come into our lovely valleys for a few days, with their fire and furnace breath, to look at the rich booty they may some day hold."[12]

Who was driving this environmental devastation? It felt like everyone and no one; it felt like it was just the way the System worked. But, as a historian, I'd point my finger at the highest levels of government and industry.

All the way back in the late 1860s, the US Army started inviting sport hunters out to the plains to kill bison, first as an anti-Indian strategy and then as a way of supplying manufacturers with leather for the belts that drove their machines. No one imagined that the vast herds, sometimes numbering in the tens of thousands, could ever face the threat of extinction. But by 1883, there were only about one thousand bison left. Many Americans were horrified to realize that modern society could wreak such destruction before anyone even had a chance to blink. It started to feel as though nature was disappearing before their eyes.[13]

Something similar happened with passenger pigeons. There were billions of them in the US in the 1870s: they were famous for congregating in massive flocks, which could darken the sky for days as they passed by. Twenty years later, only a few dozen birds remained. It turns out, if you have access to advanced weaponry and you set out to annihilate animals in the name of commercial progress (pigeon pie, anyone?), then animals tend to get annihilated. The very last passenger pigeon in the world died in the Cincinnati Zoo on September 1, 1914. She was called Martha. Which no one found funny at the time. "Millions and millions, reduced to one and now naught but history, is the sad story of the passenger pigeon," said one typical article.[14]

Maybe it makes sense that environmentally aware Americans in the Progressive era were humorless. I happen to love the Progressive movement: it gave us things like women's suffrage, food safety laws, and Tarzan. But it definitely had an elitist and moralistic streak: Progressivism was the embodiment of "white and uptight." Taken to its logical extreme, it gave us Prohibition (alcohol was considered "unsafe"—in the hands of working-class immigrants). Indeed, Progressives didn't necessarily want *all* women to vote—just the "right" kind of women. And Tarzan, for all his honor and nobility and attraction to the wilderness, was something of a sexist white supremacist. ("Me Tarzan. You Jane. You do what I say and stay away from Black Africans.")[15]

As unlikely as it might sound, one of Tarzan's forebears may have been John Muir, perhaps the most famous and best-loved environmentalist in US history (he even made it onto California's commemorative quarter). Muir, who liked to climb to the tops of trees in the midst of thunderstorms, started leading Sierra Club hikes in 1892, helping his friends to reclaim their natural vigor. Tarzan started swinging through the jungle twenty years later, in 1912. And to this day, the most potent form of environmentalism in the US draws directly on the tradition of primitivist engagement with wild, rugged landscapes.

Compare:

> Muir: "Briskly venturing and roaming, some are washing off their sins and cobweb cares . . . in all-day storms on mountains . . . ; tracing rivers to their sources, getting in touch with the nerves of Mother Earth; jumping from rock to rock, feeling the life of them, learning the songs of them, panting in whole-souled exercise, and rejoicing in deep, long-drawn breaths of pure wildness."[16]
>
> Tarzan: "Aaaaah. . . .-eeh-ah-eeh-aaaaah . . .-eeh-ah-eeh-aaaaah!" (There's no good transcription of Tarzan's call; Edgar Rice Burroughs just said it sounded like "the awful cry of the challenging ape." If you can't hear it clearly in your head, it's worth listening to Johnny Weismuller's original version from the 1932 *Tarzan* film. Weismuller, besides setting multiple Olympic swimming records, seems to have had a background in yodeling.)[17]

Anyway, over the past century, millions of nature-loving Americans have beaten their chests in national parks, thanks to Muir and Tarzan.

Muir, a classic Progressive, is pretty clearly the founding father of earnest, self-serving environmentalism. Urging Americans to drop out of the rat race for a while—and then, presumably, to go directly back to the rat race, since he never suggested any alternatives to traditional industrial development, with its rapacious hunger for natural resources—Muir insisted that his fellow citizens would thank him later if they immersed themselves in wilderness. Just hop on the train, he said (we're assuming here that you're the right kind of person and can therefore afford the price of a ticket), and go out to the Flathead Reserve in Montana—*and you will find the fountain of youth*: "Give a month at least to this precious reserve. The time will not be taken from the sum of your life. Instead of shortening, it will indefinitely lengthen it and make you truly immortal." The name of the reserve, though, suddenly reminded Muir of the Flathead Indians, so he made a point of reassuring his (white) readers that no one would dare stand between them and eternal life: "When an excursion into the woods is proposed, all sorts of dangers are imagined—snakes, bears, Indians. Yet it is far safer to wander in God's woods than to travel on black highways or to stay at home. The snake danger is so slight it is hardly worth mentioning. Bears are a peaceable people, and mind their own business. . . . As to Indians, most of them are dead or civilized into useless innocence." Muir and his compatriots were so good at erasing Native Americans from the land and from US history that today most visitors to national parks have no idea that the supposedly "pristine" wilderness they're visiting was once inhabited by thriving Native communities. We still watch documentaries referring to the national park system (institutionalized in 1916) as "America's Best Idea," without recognizing that at the heart of that idea was the killing or forced removal of numerous Indian nations.[18]

Isn't history uplifting?

(A few years ago, the writer Ta-Nehisi Coates came to my campus to give a talk arguing that the US government ought to consider paying official reparations to African Americans. He mentioned that at one point a friend had expressed concern about his emotional state: Wasn't it incredibly depressing to do all that research showing how Black people had suffered from discrimination and oppression for so many years? Coates replied, "You think I've got depressing material? Have you ever visited a History Department?")

John Muir, like so many environmentalists since, expressed a general disapproval of Civilization and its sinfulness, but he never offered a thorough critique of the system or worked through the question of his own complicity (and that of others who could afford to take a month off to walk in the woods). "The vice of over-industry and the deadly apathy of luxury" signaled to him that modernity was utterly corrupting. He had a point. He decried the way human beings were dotting the land with "clustered city shops and homes," so that no urban dweller could ever experience solitude or open space. True enough. Agriculture, it seemed to him, was most notable for having "ploughed and pastured out of existence" hundreds of square miles of wildflowers. He referred to sheep as "hoofed locusts."[19]

Meanwhile, he operated a prosperous, twenty-six-hundred-acre fruit ranch in California and frequently went on vacation to Alaska.

So it's almost impossible to imagine *not* making fun of John Muir—and I don't just mean one hundred years later. Toward the end of his life, he became infamous for opposing San Francisco's effort to build a dam in a part of Yosemite National Park called Hetch Hetchy Valley. Public support for the dam was broad, especially in the aftermath of the 1906 earthquake and fire: the city clearly needed a more reliable water supply. Muir took a stand based purely on the valley's sublimity. "Dam Hetch Hetchy!" he exclaimed. "As well dam for water-tanks the people's cathedrals and churches, for no holier temple has ever been consecrated by the heart of man." But his argument came across as selfish and elitist, because there were millions of people demanding water and hydroelectric power, compared to just a few hiking enthusiasts who seemed concerned about preserving the sanctity of the national park model. Muir took a beating in the press. James D. Phelan, former mayor of San Francisco, wondered out loud whether the founder of the Sierra Club "would sacrifice his own family for the preservation of beauty."[20]

I am not opposed to beauty, and I'm even quite fond of John Muir. I will always be grateful that the United States preserved a few slices of wilderness in the name of aesthetics and in the hope of counterbalancing the culture's dominant embrace of development, growth, conquest, and control.[21]

But why are we still stuck in Hetch Hetchy one hundred years later? The polar bears don't need more self-righteous, escapist, history-erasing wilderness preserves; they need us to get over ourselves and switch to renewables.

"We can't make the trip to Washington for the Climate Change Rally, but we can do our part from here."

Over the course of the twentieth century, as ecological interdependence became an established principle, environmentalism did develop more of a social conscience. Activists certainly continued to celebrate wildness and sublimity, but they also started talking about the universal human need for clean air and water. The idea of global interconnection, which even John Muir had an inkling of—"When we try to pick out anything by itself, we find it hitched to everything else in the Universe"—started out as a beautiful vision of common ground. Environmentalism has always had the potential to create a powerful sense of human and even cross-species solidarity. But the ethic of interdependence, alas, didn't really take hold until it became a vision of contamination. The atomic bomb, pesticides, and new industrial processes like plastic manufacturing left traces of radiation and chemical toxins in people's bodies that began to be discovered with some regularity in the 1950s. And everyone was shocked. Shocked![22]

It's sometimes tiring to be a historian. Remember how, back in the nineteenth century, people figured out that burning coal created—wait for it—coal smoke? Remember how people figured out that dumping raw sewage into the river had a tendency to kill some of the things in the river? So, yeah, it shouldn't have been surprising that releasing poisons into the environment would cause the environment to be . . . poisoned. Nature does have an amazing capacity to absorb and diffuse toxic substances, but human beings seem to have an even more amazing capacity to create and deploy toxic substances. All for the good of humanity, of course.

Anyway, as the Greatest Generation marched on and made more and more Progress injecting chemicals and nuclear waste into the soil and water, environmentalists extended their concern with the spiritual health of elite white folks to address the physical health of the entire public. This new approach exploded onto the scene in 1962, when Rachel Carson published *Silent Spring*, probably the most important environmental book ever written. Suddenly, "elixirs of death" were present in mother's milk and "the common salad bowl." Carson did a fantastic job of terrifying people, and that resulted in some immediate and important action, culminating in the first Earth Day (in 1970), the establishment of the Environmental Protection Agency (EPA; also 1970), and a wave of powerful legislation acknowledging that human health depended on environmental health.[23]

Unfortunately, if you're good at terrifying people, then people wind up terrified. And, in this case, paranoid. *Silent Spring* ushered in the new public-health approach to environmentalism, but it also reinforced the old sense of panic and tragedy. Carson put her trust in the Progressive model of muckraking journalism, of supplying the public with an exposé. It's a long American tradition: The forests are disappearing! The desert is taking over! We're running out of fuel! We're killing all the animals! The assumption was that once the science was on the table, the majority of Americans would act rationally and use their leverage to put limits on the production of toxins. And certainly some citizens were activated by *Silent Spring*. But others were further demoralized. The phrase "information overload" dates to the years immediately following *Silent Spring*.[24] And, sad to say, recent sociological studies show that when it comes to climate change, the more you know, the less likely you are to engage in activism.[25]

I teach *Silent Spring* regularly, and when I've reread it in recent years, I've asked myself the same questions I asked about most of President Obama's speeches: What if this weren't so earnest and incrementalist? What if it had delivered its bad news with a sense of humor?

Remember when the comedian Keegan-Michael Key stood beside President Obama and served as Luther, his Anger Translator? Obama made a mild statement about needing to pay more attention to climate change. Luther translated: "Hey, listen y'all. If you haven't noticed, California is *bone dry*. It look like a trailer for the new *Mad Max* movie up in there."[26]

Well, what if Rachel Carson had an Anger Translator?

RACHEL: The history of life on earth has been a history of interaction between living things and their surroundings. . . . Only within the moment of time represented by the present century has one species—man—acquired significant power to alter the nature of his world.

ROXANNE: Did you catch that? *Man*.

RACHEL: Nature has introduced great variety into the landscape, but man has displayed a passion for simplifying it.

ROXANNE: Have you ever met a bulldozer you didn't like?

RACHEL: The crusade to create a chemically sterile, insect-free world seems to have engendered a fanatic zeal on the part of many specialists and most of the so-called control agencies. On every hand there is evidence that those engaged in spraying operations exercise a ruthless power.

ROXANNE: A ruthless *male* power. A ruthless *white* male power. A ruthless white male *scientific* power.

RACHEL: This is an era of specialists, each of whom sees his own problem and is unaware of or intolerant of the larger frame into which it fits. It is also an era dominated by industry, in which the right to make a dollar at whatever cost is seldom challenged.

ROXANNE: It's a government of the one-percenters, by the one-percenters, and for the one-percenters.

RACHEL: Future generations are unlikely to condone our lack of prudent concern for the integrity of the natural world that supports all life.

ROXANNE: *What* future generations? You think you're going to be able to father any children after swimming around in DDT for a few more years?

RACHEL: The public must decide whether it wishes to continue on the present road, and it can do so only when in full possession of the facts.

ROXANNE: Take off your white coat, put down your whiskey and cigar, and *tell the goddamn truth*![27]

OK, I got carried away. In truth, I cherish the several Clean Air and Clean Water Acts of the 1960s and '70s, and it is hard to imagine a more effectual work of popular science than *Silent Spring*. But the point stands that even as environmentalism took off as a mass movement, its rhetoric remained dark, negative, judgmental, and humorless—precisely in the era now considered to represent the absolute high point of American comedy. All the other social movements of the '60s and '70s, though they could certainly get self-righteous, developed a cutting sense of humor about themselves, with the help of professional comedians who were politically engaged.[28] But environmentalists never deigned to make fun of themselves—and so they continue to be mocked ruthlessly by everyone else, and, too often, they continue to feel overwhelmed, instead of activated, by their own convictions.

OK, go ahead: ask me how many environmentalists it takes to change a compact fluorescent lightbulb.

Q: How many environmentalists does it take—

A: Hey, that's not funny! We're trying to save the planet!

American comedy started to get really good, and edgy, in the 1950s. Wait: the 1950s? The decade of drive-in restaurants, the Everly Brothers, and Howdy Doody? Well, it was also the decade of McCarthyism and film noir and bebop and Rosa Parks—not to mention comedians like Phyllis Diller and Lenny Bruce, who established stand-up comedy as a relevant and respected cultural form. In the 1930s and '40s, comedy had been just one element in vaudeville acts or variety shows. Famous mainstream comics like Jack Benny, Milton Berle, and Bob Hope were known for their precise timing and delivery, but they took their routines from standard joke books full of impersonal gags that could be replicated by anyone who happened to be invited to your cocktail party. Often, the setup involved a mother-in-law, and the punch line was followed by a rim shot. By the '50s, though, Americans were wrestling more directly with questions of conformity and commercialization, and individual personalities became much more important. Avant-garde comedians took a cue from Beat poets and jazz musicians and started offering seemingly spontaneous riffs that only they could have come up with—in part just to ensure that nobody would steal their jokes.[29]

Instead of a guy walking into a bar, it was the performer himself who had witnessed something important about his society or community or who wanted to confess his sins or flaws or quirks. Or *hers*. (This is precisely the premise of the television show *The Marvelous Mrs. Maisel*, set in the late 1950s.) "When I go to the beach wearing a bikini," Phyllis Diller used to say, "even the tide won't come in." And Diller's self-criticism, in turn, opened up the possibility of husband criticism—"his idea of a seven-course meal is a six-pack and a bologna sandwich"—which eventually led her successors to broader observations about male ridiculousness and institutionalized sexism. It was easy to make fun of mothers-in-law, but Diller was among the first people to ridicule men as a class, and she could get away with it only because self-ridicule was so central to her act. You laugh at the comic and identify with her imperfections, and that makes it easier to follow her when she trains her critical eye on the people and social structures you take for granted as solid and reasonable. *Look*, Diller seemed to say, *even a person as crazy as I am deserves classier treatment from the men in her life*. "The last time I said let's eat out, we ate in the garage."[30]

Phyllis Diller, on a good hair day

Like Diller's, Lenny Bruce's radicalism lay not just in his social satire but in his brutal honesty about himself. "You might assume," he said, "that I am an extremely moral individual, suffering pangs of conscience in an unjust society, but I am not. I am a hustler like everyone else, and will continue taking the money as long as this madness continues. Sometimes . . . I see myself as a profound, incisive wit, concerned with man's inhumanity to man. Then I stroll to the next mirror and I see a pompous, subjective ass."[31] We're definitely all subjective and probably all pompous (especially we environmentalists), and so for Bruce honesty was always supposed to lead toward empathy, and also toward the kind of self-criticism that helps people—and that means #UsToo—see their own complicity in destructive social systems and have the fortitude to reckon with it.

In one of Bruce's most famous bits, called "How to Relax Colored Friends," he enacted a conversation between a white liberal and an African American at a suburban cocktail party (the African American is the musician hired to provide entertainment). The white guy has no idea how to connect with the Black guy, so at first he makes fun of the party's Jewish host, hoping that they'll at least be able to share a casual anti-Semitism. When that doesn't work, he falls back on some comments about the Black boxer Joe Louis—a great fighter who also "knew his place." Then he feels bad and wants to get the musician some food, so he offers to look for some fried chicken and watermelon. And finally he insists on having his new friend over to his house—so long as the clearly sex-obsessed musician won't try to date his sister. "You wouldn't want no Jew doing it to your sister would you? . . . I don't want no coon doing it to my sister. No offense, you know what I mean."[32]

Alas, it's a routine that still seems timely today. Many white liberals still don't understand what it feels like to be stereotyped and would still deny their complicity in the perpetuation of stereotypes (not to mention climate change). But Bruce took his material even further, pointing out that a lot of good liberals in the North avoided confronting their own racism by blaming all the ills of segregation on white southerners. From Bruce's perspective, even KKK leaders, or Nazi leaders for that matter, deserved some empathy, since they were just doing what was normal within their social system. "The liberals are so liberal," Bruce said, "they can't understand the bigots."[33] Everybody is guilty, because human beings are corrupt. "Of course, I'm corrupt, too. If I wasn't I'd pick up your tab."[34] But everybody is also redeemable: you just have to be willing to see yourself clearly and start addressing your hypocrisy. Even the cluelessly vicious white guy at the cocktail party needs to be part of the revolution. "We assume that this cat is all bad . . . , but there are sensitive parts to him also, man."[35] First step: admit you're a bastard. Second step: try to be a little bit less of a bastard. There's just as much room in Purgatory as there is in Hell—probably more.

Not many white comedians were as willing as Bruce to confront racism and Jim Crow, but when the civil rights movement gained momentum in the early '60s, several African Americans got off the so-called Chitlin' Circuit of all-Black nightclubs and found new opportunities to address white audiences. Redd Foxx, Slappy White, Bill Cosby, Godfrey Cambridge, Nipsey Russell, Moms Mabley, Flip Wilson, and especially Dick Gregory started to ask more publicly why integration was so difficult in the US. Answer: because white people are bastards. And a lot of the most powerful white Americans are descended from slaveholders. In the end, African American comedians made an absolutely crucial contribution to the success and high spirits of the civil rights movement.[36]

Like Diller and Bruce, Gregory recognized that a good first step is always self-ridicule: "I've got to make jokes about myself before

Dick Gregory, the consummate comedian-activist

I can make jokes about them and their society—that way they can't hate me."[37] And that way I don't hate myself, either: I see myself honestly, without getting depressed about it. Environmentalists, take heed!

Gregory was scathing in his attacks on white power—he quickly went from being a comedian to being a civil rights leader—but first he drew his audience in by acknowledging his own weaknesses and *almost* playing to the crowd's stereotypes. Puffing on a cigarette, he admitted, "I been readin' so much about cigarettes and cancer, I quit readin'." So maybe the white folks in the audience could peg him as just another uneducated addict? Then again, maybe he had more in common with them than they realized. "One night I got so drunk," he liked to say, "I moved out of my own neighborhood." Pause for nervous laughter. And if that joke didn't register fully, Gregory had a follow-up that addressed white flight and housing discrimination a little more directly: "Lookin' for a house can be quite an experience. Especially when you go into a white neighborhood, offer forty thousand dollars for a twenty-three-thousand-dollar house, and then get turned down 'cause you're lowerin' the property values."[38]

After Martin Luther King Jr. was assassinated in 1968, Gregory gave up his career as a professional comedian—but he didn't give up comedy. Hired on the campus lecture circuit as a civil rights speaker, he would start each of his talks with forty-five minutes of stand-up. (Then he would start yelling, "America is the number one racist country in the world!" Whoo-hoo!)[39] It makes sense to shout sometimes, but some of the most difficult messages reach their target more easily when they're delivered as jokes. One night in Chicago, Gregory learned suddenly that his audience was composed mostly of white southerners, executives in the frozen-food industry. "I know the South very well," he asserted.

> I spent twenty years there one night. . . . I walked into this restaurant. This white waitress came up to me and said, 'We don't serve colored people here.' I said, 'That's all right, I don't eat colored people, no way! Bring me a whole fried chicken.' About that time, these three cousins came in. You know the ones I mean—Ku, Klux, and Klan. They said, 'Boy, we're givin' you fair warnin'. Anything you do to that chicken, we're gonna do to you. . . .' So I put down my knife and fork, picked up that chicken, and kissed it![40]

Perhaps more than anyone else in the history of US popular culture, Gregory opened up space in the mainstream for serious social criticism. Sure, many white northerners were converted to the civil rights cause in the 1960s by photographs and footage of police brutality against unarmed Black protestors, by images of fire hoses and nightsticks. But many were also moved by Gregory's comedy shows and albums. Sometimes he was polite, mellow, a bemused observer of humanity's foibles: "My daughter, she doesn't believe in Santa Claus. She knows doggone well no white man is coming into a colored neighborhood after midnight." At other times, he was acerbic: "You know the definition of a southern moderate? That's a cat that'll lynch you from a low tree."[41] Gallows humor! Black comedy! It worked. Hundreds of comedians—and activists—have followed his lead, from all backgrounds and orientations. Over the years, curmudgeonly critics have tried to dismiss Gregory's brand of humor as simultaneously too political and not political enough: either the jokes were ruined by his serious intent to convey a message, or the seriousness of his message was undermined by the jokiness of his delivery. But the recordings of his early-to-mid-'60s shows in front of white audiences suggest that they were with him all the way. By artfully demonstrating his own humanity, he spurred his listeners to get back in touch with theirs.[42]

In the '70s, Gregory's most obvious successor was Richard Pryor, and he went after not only complacent white people in general but also the group of liberals he thought might have been the most hypocritical of all: white environmentalists. The environmental movement had exploded between the publication of *Silent Spring* in 1962 and the passage of the 1972 Clean Water Act, but all those Greens were still rather pale. It was hard to escape the feeling that people's concern "for the planet" was mostly a concern for their own neighborhood, and the better-off, white neighborhoods seemed to be the only ones getting cleaned up. Pryor was especially put off by white advocates for population control, as he explained to an integrated audience in a 1975 bit called "Shortage of White People" (I've tried to make the language more family friendly, but you could listen to a recording for the full effect): "I ain't seein' no white folks no more. Y'all stop [having sex]? . . . They stopped [having sex] because some rich white man told 'em, said [*putting on a white voice*], 'Look, come on, cut the crap. Jesus Christ, there are too many people on Earth! I have no place to ride my horsy!' [*Shifts back to his own voice*] There will be no shortage of [Black folks]. [Black folks] is [having sex]. We got to have somebody here to take over!"[43]

Pryor took the age-old stereotype of oversexed Black people and somehow used it to make nature-loving white folks seem utterly unnatural. Population control was just another power play by privileged, pleasure-seeking elitists who wanted to keep all power, privilege, and pleasure for themselves. At a time when every other social movement sought to widen the circle of humanity, environmentalists had devised a program that seemed antihuman. So Pryor and his audience laughed at them. If Freud was even half right about anything (no disrespect to the asexual community), then Pryor's joke should have offered the ultimate in common ground and solidarity: almost everybody wants to have sex, right? Not environmentalists, apparently.

And environmentalists also hate babies. More recently, an entirely serious green advice columnist ended her screed about sustainable birth-control methods by noting that "no matter what type you choose, it's guaranteed to have less of an impact on the environment than the unwitting creation of a fossil-fuel burning, diaper-wearing copy of yourself."[44]

"Ouch," said the father of three very hungry boys. (At least I got a vasectomy after number three. And my wife and I are raising the kids to be tree-hugging vegan bicyclists. But we do sometimes eat local meat. And we occasionally drive somewhere to go hiking. Oh, screw it. I'm going back to my rock garden.)

In the end, I prefer the population-control method proposed by the comedian Bill Burr. He explained to Conan O'Brien that "there's nothin' wrong with drivin' a gas-guzzling car; there's just too many people doin' it." (You can take "doin' it" in two senses.) "So we gotta figure out a way to, like, thin out the herd."

> Q (Conan): OK, how?
>
> A (Bill): I would randomly sink cruise ships. . . . I think it's a good mix of people to get rid of.[45]

And this was *before* the COVID-19 pandemic ravaged the commercial cruising fleet! Burr seems almost like a prophet now.

Back in the '70s, the population-control movement was pretty clearly not targeting the kinds of people who could afford cruises. It was the incredibly fertile (and poor) Brown and Black people who needed lessons in abstention. So Pryor was absolutely right to ridicule advocates of population control: their program, when linked with ongoing environmentalist appeals to save beautiful wilderness areas and charismatic animals, added up to the familiar elitist (and imperialist) lament that the masses were starting to encroach on various kinds of gated communities. Don't let those Mexicans near the Sierras![46]

Despite the new environmentalist concern for public health, then, the movement still struggled to articulate an inclusive positive vision. Consider the now-classic environmentalist story *The Lorax*, published in 1971 (and made into a blockbuster movie in 2012, because that's how little our environmental thinking evolved in forty years).[47] In a way, the book was a rule-proving exception, because Dr. Seuss, while clearly harboring green sympathies, actually poked some fun at the environmental movement. The title character, who serves as the mouthpiece for all the world's trees, speaks in a voice described as "sharpish and bossy," and he is in fact a nattering nabob of negativism. In one climactic scene, the book's nominal villain, the Once-ler (who uses things only once before discarding them), confronts the Lorax and basically asks him if humanity has any place at all in his version of the world. "All you do," the Once-ler notes, "is yap-yap and say, 'Bad! Bad! Bad! Bad!'" In other words, the Lorax is an annoying little dog, yipping away in agitation but never spelling out a plan for human beings to live in peace with the ecosystems they inhabit.

This acknowledgment of environmentalism's shortcomings is part of what makes *The Lorax* an effective book, though Dr. Seuss's self-ridicule tends to get forgotten in the rush to celebrate his prophetic concern for ecosystems. And Dr. Seuss did effectively incorporate the latest ecological research by pointing out the interdependence of the forests and ponds and mammals and birds and fish: when you chop down all the trees to feed factories that cater only to consumerist "needs," and when those factories belch out such "by-products" as "Gluppity-Glupp" and "Schloppity-Schlopp," then the whole area will go to Hell, and no animal of any kind will be able to live there.

The impression we're left with, though, is that what matters most in the book is precisely the Truffula Trees, the "rippulous" ponds, the Brown Bar-ba-loots, the Swomee-Swans, and the Hummingfish. Where are the people? The only well-developed character in the story is the Once-ler, who seems to represent all of humanity and thus to suggest that human beings are irredeemably stupid and greedy. There once was a paradise; then people arrived and ruined it. Bummer! What do we do now? The Once-ler, in his dotage, has regrets, and at the end of the book, he offers the very last Truffula seed to the reader, who is given the humble task of saving the world. There's no hint of social structures or power dynamics, no suggestion that corporate privilege needs to be reined in by government regulation—or popular activism. Just an invocation of personal responsibility and a hope that you'll go out and plant a seed somewhere (where?), so as to create a new, pristine forest, to be inhabited by the Lorax and his animal friends. And then, maybe, if you're sophisticated enough, you'll be able to ride your horsy there!

Alas, as the environmental movement grew in the '70s and into the '80s and '90s, its leaders continued to harp on issues that made it seem as though they hated all human beings. (One might reasonably ask, Did they also hate themselves? The Jews and Catholics probably did, but they were in the minority.) When environmentalists insisted that we save "old-growth forests" or "spotted owls" or "the world's greatest wild salmon run," they were explicitly excluding human beings from their definition of "the environment." And so their opponents could legitimately ask, What about the people who depend on forest products for their livelihood? Are owls more important than people? Can we still eat bagels and lox?[48]

In a 1999 episode from the first season of *West Wing*, three activists try to convince White House press secretary C. J. Cregg that the new, progressive administration should build an eighteen-hundred-mile freeway running from Yellowstone to the Yukon—a freeway reserved for wolves.[49] At first, Cregg, played by Allison Janney, tries to lighten the mood by interjecting a joke whenever one of the advocates pauses for dramatic effect, but her attempts at humor are met with either confused silence or prickly defensiveness.

> Q (Cregg): How are you going to teach wolves to follow road signs?
>
> A: Our scientists are working on a plan.

Eventually, Cregg decides she's heard enough, because it becomes clear that the environmentalists care far more about wolves than people. When she asks about the ranchers whose livestock might be eaten by the wolves, the response is quick: "Ranchers are killers." Actually, Cregg says, "ranchers face the following conditions: falling stock prices, rising taxes, prolonged drought, and a country that's eating less beef." In other words, they're human beings, with understandable motivations. Plus, they vote! Wolves don't!

When the activists desperately point out that the cost to the taxpayers will be "only $900 million," Cregg finally just collapses in laughter. She thinks they *must* be joking this time. They're not. Environmentalists never joke.

It might be their humorlessness, more than anything else, that has made it so hard for environmentalists to communicate with the public about climate change. If we Purveyors of Gloom just acknowledged our own antihuman bias every now and then, we might seem a lot more, well, human.

Once you start making fun of yourself, you can say almost anything, and it will sound a lot less preachy. You could even say that we should pay less attention to wolves and polar bears and the Approaching Apocalypse and instead try to protect the vulnerable people currently struggling to survive in polluted, flooded, burned-out, or drought-ridden neighborhoods and regions, whose problems are increasingly the result of our addiction to fossil fuels.

Or, if you're working at *The Onion*, you could just put out a list of "Tips for Combating Climate Change":[50]

—Minimize your carbon footprint by always buying locally sourced petroleum.
—Inhale smog directly from any nearby exhaust pipes to prevent it from getting into the atmosphere.
—Realize that you have the power to make a difference—just not enough to divert humanity from its catastrophic course.

Allow me to translate. Environmentalists! We need to get over ourselves!

Or, alternatively: Look, we don't even have to give up the panicked alarmism, as long as we take ourselves a little less seriously!

An acquaintance of mine from the 1990s DC environmental scene, Chip Giller, moved to the West Coast and started a green website called *Grist*, which has offered itself for the past twenty-five years as "a beacon in the smog," delivering "doom and gloom with a sense of humor." In the late twenty-teens, *Grist* even started hosting an annual Comedy Extravaganza in New York City. From the very beginning, Chip committed himself to making fun of environmentalists' age-old tendency toward dreariness, which has allowed him to be a much more effective environmentalist, even when he's felt obliged to lecture us about our carbon footprint.

And Colin Beavan's 2009 comedy *No Impact Man* is one of the best environmental films ever made precisely because it's honest about what a demanding, self-important, extremist prick Colin Beavan is. When Beavan's wife, Michelle Conlin, starts to complain about him—If it's No Impact *Man*, she suggests, then why does the whole family have to give up electricity?—we can understand her frustration. We empathize as Michelle mourns the loss of her television and starts writing on the walls of the apartment with bright-colored magic markers, and as she threatens to murder Colin if he makes her go another day without coffee, and as she asks him if maybe he's become a "fringe wacko." The whole movie is designed to keep us in a good mood by gently making fun of the person trying to force sacrifices on us for the sake of the climate. But of course we know that Colin is the one giving Michelle all that screen time, so he becomes charming in his self-awareness, and ultimately we want the marriage to work out, so we think, "OK, No Impact Man, in exchange for the fun of seeing you ruthlessly mocked, we'll actually listen when you start talking about droughts and floods and epidemics and refugees. We'll even meet you halfway and start using a clothesline. Maybe we'll even walk to the farmer's market instead of driving to the grocery store. You might even convince us to attend a rally for climate justice."[51]

But we're keeping the damn television, so that we can rewatch your movie, you pontificating, doom-saying, people-hating wing nut.

2

PURGATORY

Perpetually Improvising on the Edge of the Abyss

(Or, What Environmentalists Can Learn from Jews and African Americans about Gallows Humor)

Chris Rock has had some adventures at the Oscars. Things got particularly hairy (pun intended) in 2022 when he drew attention to Jada Pinkett Smith's baldness and immediately got slapped by her husband. But I think Rock was in an even tighter spot when he hosted the show back in 2016.

That was the year—OK, one of the years—when there were no African American nominees in any significant category, and a number of Black celebrities called on Rock to join their boycott of Hollywood's biggest party and step down as host. But he demurred. In his opening monologue, he invoked history to explain his decision. Why are we protesting this particular Oscars ceremony, he wanted to know, when there have been no Black nominees at least seventy-one other times? Take the 1950s and '60s: if Sidney Poitier didn't put out a movie, there were probably no Black nominees. But African Americans weren't boycotting the Academy Awards in the early 1960s because, as Rock put it, "we had *real* things to protest at the time." Huge round of applause: yes, all of white, liberal Hollywood can agree that the civil rights movement was more important than who gets recognized for their contributions to the film industry. That was the safe joke to make. But successful comedians usually push the limits of appropriateness: they want to connect with their audience, want to earn some empathy, but only if the audience is willing to be affronted and unsettled, to join in the violation of taboos, to see things differently. So Rock raised the stakes: "We were too busy being raped and lynched to care about who won best cinematographer." Scattered applause; a few laughs. The audience was still with him, but now they were on edge, which is right where Rock wanted them: "When your grandmother's swinging from a tree, it's really hard to care about Best Documentary Foreign Short."[1]

Dick Gregory would have been proud: African American comics have a solid tradition of converting the tragic history of lynching into gallows-humor counterpunches. (For the record, "Best Documentary Foreign Short" is not a real award category.)

Rock's last line in that bit got very little overt approval, but it had a huge impact. Usually you want raucous laughter, but stunned silence can also indicate that your jab has landed. Rock had managed to make fun of everybody. He had suggested that Black celebrities like Spike Lee—and Jada Pinkett Smith—might be getting a little too exercised about something as trivial as the Oscars. But at least Spike and Jada, he seemed to acknowledge, were not as complicit as his mostly white audience in the kind of racism, past and present, that actually seeks to destroy Black bodies and spirits. In early 2016, the entire country was aware of the Black Lives Matter movement and the many recent cases of police brutality against African Americans—which is why, another minute into his monologue, Rock noted that this year the "In Memoriam" montage at the Oscars was going to be different:

instead of honoring departed movie stars, it would be devoted entirely to "Black people that were shot by the cops on their way to the movies."[2]

Now the applause was even more scattered. And the laughter sounded kind of confused. *What did he just say?*

What he said was, *I'm in a bit of a tight spot here*. His job was to be funny and entertaining. You can't be sullenly political at an awards show. When Rock made fun of Jude Law at the 2005 Oscars, and Sean Penn defended Law without even a hint of a sense of humor, everybody made fun of Sean Penn the next morning. On the other hand, Rock wasn't asked back to host the Oscars for eleven years. And now his Black friends thought he was betraying the cause. Meanwhile, his white audience would have preferred to be serenaded by Billy Crystal or Neil Patrick Harris. But Rock had every intention of getting paid for this gig. Ultimately, he decided to expand on the "Oscars So White" controversy to make some really serious points about the violence of racism in the US, but he delivered those points in the form of jokes about how all celebrities, Black and white, are ridiculous, self-congratulating narcissists. (Go ahead, celebrities, name one cinematographer or one award-winning documentary short.)

Again, why haven't environmentalists ever figured out how to walk this line?

Last I checked, Chris Rock had not solved the United States' racial problems. But at the 2017 Oscars, the Best Supporting Actor and Actress awards both went to African Americans (Mahershala Ali and Viola Davis). And the Best Picture award—though it was first mistakenly presented to *La La Land*, which felt like just another expression of the Academy's self-satisfaction, since that film was in many ways an homage to Hollywood—

ultimately went to *Moonlight*, the only Best Picture winner ever with an all–African American cast. The entire Oscars So White and Black Lives Matter movements deserve credit, but Rock's impact was especially clear, since he was serving as the face of the Oscars and he had 34.4 million viewers.[3] Of course, shifts in Hollywood politics are never going to compensate for hiring and housing discrimination or voter suppression or police brutality: that was part of Rock's message. But another part of his message was, *You better take these jokes seriously and make some changes.* It was as if the Academy had finally invested in renewable energy sources, which were now getting to compete with fossil fuels on something closer to a level playing field. Turns out renewables are pretty darn good, as their advocates have been saying for decades.

Of course, Rock had the chance to write and practice his monologue before the Oscars broadcast, but I still think of it as an improvisation of sorts, because all comedians make adjustments on the spot, and, more important, back when he accepted the gig, he had no way of anticipating how controversial his appearance would become. Fortunately, because so much of comedy, especially dark comedy, invokes the spirit of improvisation, it's the perfect genre to help you adapt to situations that you'd probably prefer not even to acknowledge. And adaptation beats the heck out of mere acceptance when it comes to social issues.

If I want to accept my personal fate—middle age, back pain, academia—I go see *Oedipus Rex* or *King Lear*. But if I want to think about shifting a few things around in the world, in spite of the inevitability of Republicans, I go see an improv comedy show. The whole mechanism of improvisation relies on your ability to respond to rapidly evolving circumstances. That's what we need right now. An improv attitude is the only thing that's going to get us through these dark times.[4]

But, you may ask, what's the point of improvising right before you meet your doom? Why make jokes on the gallows? Dark comedy often gets a bad rap: people say it's too fatalistic or even nihilistic. It's not like the hangman is going to think you're so funny that he'll decide not to pull the lever.[5]

On the other hand, though, why go to your death in a bad mood? At the end of Monty Python's *Life of Brian*, the two-dozen men hung up on crucifixes in the desert suddenly start singing a cheerful ditty: "Always Look on the Bright Side of Life."[6] It's like a twisted version of "My Favorite Things" from *The Sound of Music,* or a response to all those Disney toe-tappers reassuring kids that it's going to be OK. No, it's not going to be OK: sometimes, when the dog bites or the bee stings, you wind up in anaphylactic shock, and not even brown paper packages tied up with strings will bring you back. But for centuries philosophers have been urging us toward better, nobler ways of dying, and what could be better than singing yourself offstage? Why give your executioner the last laugh? You can die with a look of despair, terror, and defeat, or you can give your audience a wink and make fun of the hangman's ridiculous hood.

Gallows humor should be second nature (as it were) to environmentalists. What could be more natural than death? Accepting death is the most basic way of upholding limits, of acknowledging the finitude of our planet's resources.[7] Thank Heaven people die (and also decompose), because if they didn't, we would have run out of space a long time ago.

In the end (or just before the end), gallows humor and dark comedy are basic human coping strategies. They make Purgatory more bearable for however long we're in it. But their improvisational mood—that sense that you're creating the tiniest bit of wiggle room in a hopelessly cramped space—can also help you take aim at the people who are cramping your style. Thus the organic link between gallows humor and satire: if you ridicule your oppressors cleverly enough, you might be able to siphon off some of their power while you're on your way out, and that can be helpful to everyone who's left behind. The first scholar I know of who studied gallows humor, Antonin Obrdlik, asserted in 1942 that it should be understood as "an index of strength or morale on the part of oppressed peoples." That's right, 1942: his argument was based largely on "experiences in Czechoslovakia following the advent of Hitler."[8]

The problem of climate change is similar to the problem of racial and ethnic injustice in that we're all guilty, to some extent. But we also need to remember that some people are more guilty than others. American Petroleum Institute: we're coming for you. Why haven't we been doing black comedy about coal and oil companies for the past several decades?

Purgatory demands dark comedy. Imagine that you're stuck in a waiting room, seemingly forever. It will at least help your mood to improvise some riffs on the unbearable 1970s wallpaper. But if you can get the receptionist on your side by also making fun of the bosses who established the rules of the waiting room and hung that wallpaper in the first place, then you're one step closer to freedom. Or the *possibility* of freedom. It's Purgatory, after all.

Some literary scholars have argued that black comedy and gallows humor are ultramodern phenomena; some have even insisted that they didn't really exist in their pure form until the mid- to late twentieth century, after Hiroshima, Nagasaki, the Holocaust, and the first few seasons of the New York Mets.[9] But—need I remind you?—I'm a historian, so I'm going to propose that they've been around somewhat longer. I believe that human beings have been improvising and laughing in the face of horror for at least several centuries. Just ask the African Americans and the Jews. Or the Irish or Native Americans or women or the queer community or countless other oppressed people. But I think the long traditions of African American and Jewish humor have the most to offer us in our current fight against climate change.

Black comedy for Black people by Black people goes back at least to the time of slavery, though I imagine that certain forms of African wildlife inspired some gallows humor in even earlier periods. For enslaved people, maintaining a sense of humor was clearly a survival strategy; it could also generate creativity and even resistance. Of course, the particular style of comedy depended on the company: "Got one mind for the white folk to see, / 'Nother for what I know is me."[10] But almost all of the Black humor that arose during slavery expressed, either directly or indirectly, the exceedingly relevant paradox that slavery was both a fact that one had to live with and an egregious crime that had to be overturned. Climate change, anyone?[11]

It's perhaps too simplistic to say that the ruling classes in the West replaced their dependence on slavery with a dependence on fossil fuels, but, honestly, doesn't a lot of modern history pivot around the gradually increasing laziness of rich white people and their quest for more and more elaborate ways of luxuriating?[12]

Of course, the ultimate form of luxuriating is going to Heaven, an idea ripe for comedic exploitation—as in this example of African American humor from the slavery era, transcribed by the Harlem Renaissance poets Langston Hughes and Arna Bontemps:

> One morning, when Ike entered the master's room to clean it, he found the master just preparing to get out of bed. "Ike," he said, "I certainly did have a strange dream last night."
>
> "Sez yuh did, Massa, sez yuh did?" answered Ike. "Lemme hyeah it."
>
> "All right," replied the master. "It was like this: I dreamed I went to [Black] Heaven last night, and saw there a lot of garbage, some old torn-down houses, a few old broken-down, rotten fences, the muddiest, sloppiest streets I ever saw, and a big bunch of ragged, dirty Negroes walking around."
>
> "Umph, umph, Massa," said Ike. "Yuh sho' musta et de same t'ing Ah did las' night, 'cause Ah dreamed Ah went up ter de white man's paradise, an' de streets wuz all ob gol' and silvah, and dey wuz lots o' milk an' honey dere an' putty pearly gates, but dey wuzn't uh soul in de whole place."[13]

Ike acknowledges that his master can say whatever he wants to, that he can engage in casual calumny from the moment he wakes up, that he can make Ike's servitude a living hell. But with one quickly improvised joke, Ike manages to equalize their relationship: here we are, together, in Purgatory. You're never going to make it to Paradise, Massa, because all white people are complicit in the moral turpitude of the slave system. And in this version of the story, clearly intended for an African American audience, the generic, nameless master can't even justify punishing Ike for his joke's content, because it was the creation of Ike's subconscious. A man is not responsible for his dreams, after all. Of course, masters did not *have* to justify the punishments they meted out to enslaved people, so Ike was clearly taking a risk. But the emphasis on his clever use of irony, on his rhetorical triumph, on his ability to outsmart his master even when put on the defensive, made this a story capable of boosting African American morale over the course of many decades, despite unceasing affronts and indignations.

Speaking of affronts and indignations, remember when Senator Jim Inhofe (R-OK), chair of the Environment Committee, brought a snowball into the Senate in 2015 to prove that global warming was a hoax? Well, I wish Ike had been there. He might have looked around and said, "Well, I'll be damned. A snowball in Hell."[14]

Ike is just one of countless characters from slavery times who specialized in gaining some purchase on miserable, intractable situations. A figure called Old John survived in African American culture well into the twentieth century. After the writer Zora Neale Hurston moved down to Florida in the late 1920s to do some anthropological field work, she often sponsored what she called "lying contests," and she found that many of the storytellers, drunk on moonshine, wanted to tell her about the triumphs of Old John.

There was the time, for instance, when John's master bet a considerable sum of money that John could beat an enslaved man from his neighbor's plantation in a fistfight. But when they arrive at the appointed spot, they find that John's opponent is a giant, pulling hard against a chain staked to the ground, flailing his arms and bellowing like a wild beast. John sizes up the situation and calmly walks over to his master's wife, whom he slaps in the face. At first his master is furious, but when the opponent turns tail and runs away so energetically that he pulls out the stake, John's master wins the bet by forfeit, and all is forgiven. John ultimately explains that he knew the other man would be terrified of him as soon as he saw that John had the gumption to assault a white woman.

Of course, that kind of assault is another huge risk for an enslaved person to take, and in a different Old John story, his master decides that he's had enough of John's shenanigans and tells him he's going to hang him from one of the plantation's tallest trees. So John tells a friend to climb the tree and hide near the top with a box of matches. That night, John's master takes him to the tree and tells him he can say one last prayer before he breathes his last. John falls to his knees, and says, "Oh Lord . . . if you're gointer destroy Ole Massa tonight, with his wife and chillun and everything he got, lemme see it lightnin'!" Then John's friend lights a match. John looks up at his master, who stares right back and tells him to quit praying. But John repeats the prayer two more times, and each time it's answered by a flash of light, and at that point, the master can't take any more and runs away. As one of Hurston's storytellers put it, Ole Massa "run so fast that it took a express train running at the rate of ninety miles an hour and six months to bring him back, and that's how [Black folks] got they freedom today."[15]

As a historian, I feel obliged to note the anachronism of that conclusion (trains could not go ninety miles an hour back then), but of course the point is that John, thanks to his clever improvisation, was able to overturn a power differential that had made his doom appear certain.

(We historians should have learned by now to stop fretting about anachronisms. A few years ago, when a friend and I went to see the movie *Django Unchained*, which is set in the late 1850s, I started getting all hot and bothered because a bunch of white vigilantes had donned hoods and were riding over some hills at night, brandishing torches and whooping about the lynching they were getting ready to perform. I wiggled in my seat for a few seconds and then leaned over to my friend, an innocent ecologist, and whispered, "The KKK didn't start until 1866—*after* the Civil War. It was a reaction to the Thirteenth Amendment." My friend nodded, grateful for the historical perspective. And then our attention turned back to the screen, and we saw that the riders had stopped for a quick consultation about how to proceed with their reprehensible business. And one by one, they started ripping off their hoods, complaining about how uncomfortable they were, how hard it was to see out of them, how faulty the design was. It was a hilarious bit, which I laughed at even harder than I normally would have because I was so embarrassed about second-guessing a comedy. Quentin Tarantino, 1; Know-It-All Historian, 0.)

Anyway, Old John represents a classic figure in African and African American folklore, someone we environmentalists might want to get to know: the Trickster. John is a direct descendant of Eshu, who is understood among the Yoruba people of Nigeria as, on the one hand, "the enforcer of sacrifice," and, on the other, an expert diviner and master of fortune reversal. Basically, if you're willing to pay, Eshu might have a way of sparking a happy accident on your behalf. In traditional Yoruba stories, Eshu sometimes has to remind people that they are in fact caught in binding webs of destiny, but he also thrives on contradiction and is eager to demonstrate that life is unpredictable. His is a complicated but useful attitude, which might be summarized as follows: we're probably screwed; but what if we were to take some risks and try to tweak the system a little bit?[16]

Q: Hey, Eshu, what are the odds that the next hurricane will hit my house?

A: Well, that depends; if you turn down your thermostat and pay me the money you saved on your heating bill, and also walk up to Melania Trump and slap her in the face, then I'll see what I can do about turning the next hurricane away from your house and toward Mar-a-Lago.

Eshu is on to something: if you're going to ask people to sacrifice, it's best to have a sense of humor about it. In the summer of 1979, President Jimmy Carter made his infamously earnest, pained "crisis of confidence" speech, asking people to carpool and ration their gas and obey the speed limit. Within a few months, Ronald Reagan was mocking the president ruthlessly on the campaign trail, asking voters if they preferred his smiling optimism or Carter's "malaise."[17]

Maybe President Carter should have had George Carlin give the speech instead. Comedians are practiced in the art of affronting their fans in an almost endearing way: the goal is often to be "playfully mean" to the audience.[18]

Here's President Carter's attack on the culture of consumerism: "Too many of us now tend to worship self-indulgence and consumption. Human identity is no longer defined by what one does, but by what one owns. But we've discovered that owning things and consuming things does not satisfy our longing for meaning. We've learned that piling up material goods cannot fill the emptiness of lives which have no confidence or purpose." The president starts with selfishness, builds to a climax of meaninglessness and emptiness, and leaves us crying tears of hopelessness into a puddle of purposelessness. Vote for Carter![19]

Compare Carlin's simple assertion that Americans must be "dumber than a second coat of paint. . . . Only a really low-IQ population could have taken this beautiful continent, this magnificent American landscape, that we inherited—well, actually we stole it from the Mexicans and the Indians, but it was nice when we stole it, looked pretty good; it was pristine, paradise— . . . only a nation of unenlightened half-wits could have taken this beautiful place and turned it into what it is today: a shopping mall." Now, I'm not claiming that laughing at your complicity in colonialism and runaway consumerism is necessarily going to inspire you to do the right thing, but you have to agree that it's better than listening to Jimmy Carter commit political suicide.[20]

Alternatively, you can sometimes take serious exhortations out of their old context and deploy them in ways that might make them more palatable. Enough activists and artists have revived this pro-carpooling poster from World War II that it's become a meme.[21]

In African American culture, Eshu the Trickster sometimes became the Signifying Monkey—another compelling and potentially useful character. Tricksters can often transcend simple binary oppositions, like the master-slave relationship or, in the kingdom of nonhuman animals, the predator-prey relationship. A monkey will always be hunted by a lion, but if he learns to "signify," that is, to use all the tricks of language, he might be able to turn the lion's aggression elsewhere. The classic Signifying Monkey sits up on a tree branch basically hurling a succession of "yo' mama" jokes at Lion—it's just that he casually attributes all of them to Elephant. After a while, Lion starts roaring and pacing and finally rushes off to pick a fight with Elephant, who casually beats him to a pulp. It's the triumph of mental dexterity, of improvisational wit, over power and authority. And the stakes get even higher when the King of the Jungle comes limping back, and Monkey starts gloating, hopping up and down on his branch in glee—until suddenly he slips and falls, and Lion pins him to the ground. Now Monkey *really* has to start signifying. He apologizes—but instead of simply begging for his life, he proposes to give Lion an even more thorough beating, if the King will just step off and give him a fair chance. Lion licks his chops and loosens his grip, and Monkey jumps right back into the tree, immediately reverting to his monologue of mockery and vilification.[22]

Basically, the King of the Jungle is a loser. He's all aggressive instinct and arrogant confidence. He's like a slave master or a captain of industry who assumes that what he's doing is natural and right, that he's won the competition and so deserves his spot at the top of the food chain. He's a one-trick feline.

But if you can signify, then you can adapt to the trickiest situations. You're versatile and flexible, like Ike and Old John, slyly undercutting their masters with false flattery and double entendres and innocent-seeming jabs. You can employ culture to trump nature. You can fake it till you make it.

Aren't you sick of the hopeless debates between environmentalists and climate-change deniers? Between Democrats and Republicans? Between people of integrity and billionaire politicians who are in the pockets of billionaire tycoons? You're not going to beat Senator Inhofe in a regular fight. You have to fake him out, get a little distance from the game, find a way of hopping up onto a branch and pissing down on his snowballs.

In some Native American cultures, the Trickster Coyote plays the role of the Signifying Monkey. Coyote is adored for his cleverness and sense of humor but also scorned, because he brings people no end of trouble. Sometimes, people set traps for him, especially by setting out poisoned animal carcasses, trying to take advantage of his nature as a scavenger. The carcass occasionally attracts and kills an entire band of wolves. But the solitary coyote almost always sniffs out the trick: he would gladly go against his nature, gladly restrain himself, gladly sacrifice a meal, for the sake of survival. And he always leaves a little improvised message for the trappers, a critique of their primitive techniques. *I'm out of your league*, Coyote says: *I pee and poop on your pathetic traps.*[23]

Playing the Trickster can help you not only confront your enemies but also rouse your allies, as President Carter was trying to do (but without any sense of humor). More recently, President Obama found himself with a similar goal when he finally decided in his second term that he needed to gain some traction on climate change. He also finally decided, as a lame duck, that he could abandon his lame gradualism and loosen up a little: after the midterm elections of 2014, when asked if he had a bucket list for the final two years of his presidency, Obama said, "No, but I have something that rhymes with 'bucket list.'" So, at a Democratic fund-raiser the next year, the president referred explicitly to the Senate Snowball Incident and then posited, with the help of a perfect analogy, that you'd have to be totally crazy not to want to do something about climate change. OK, he says: you're not feeling well; so "you go to one hundred doctors, and ninety-nine of them tell you, 'You've got diabetes. You've got to stop eating bacon and donuts every day. . . .' You wouldn't say, 'Awwww, that's a conspiracy. They're making that up. All ninety-nine of those doctors got together, with Obama, to try to prevent me from having bacon and donuts.' You wouldn't do that!" He might have added that the one doctor who wanted to let you keep eating bacon and donuts every day was probably paid off by the bacon and donut lobby.[24]

President Obama's blunt, pragmatic, fact-facing kind of humor, that smiling realism in response to yet another ridiculous outrage, aligns him not only with a long and long-suffering African American tradition but also with an even longer Jewish tradition. African Americans and Jews are really good at using humor to shake people out of their depression and inertia. And you can especially count on the Jews when the latest outrage threatens to explode into the Apocalypse.[25]

I grew up hearing a joke about a modern Noah-style flood that was going to put all the world's humans out of their misery. So the pope goes on TV and says, "Your local priest will be available for confession twenty-four hours a day until the water starts to rise." The head minister comes on and says, "Please go see your local pastor right away for family counseling and prayer." And then the chief rabbi says, "Listen, I want everyone to get down to the Jewish Community Center immediately: today only, we're offering *free swimming lessons*!"

I know a number of Reform Jews who would never give up bacon, no matter what their diagnosis, but when it comes to the end of the world, we're ready to improvise.

Pretty much all Jewish humor is gallows humor. It can be summed up in one word: "Oy!" All the kvetching we do is just a way of acknowledging the unceasing parade of horrors that constitutes daily life on this planet.[26]

Think about what we know of early Jewish history. Yes, the Lord parts the waters of the Red Sea, allowing the Jews to scramble across and escape. Pharaoh's troops get drowned. But then what happens? The Jews are almost immediately lost in the desert, where they wander for the next forty years. A few weeks into their exile, God offers them the Torah, which is basically a list of rules, commands, and restrictions on freedom: no bacon, no shrimp, no foreskin. The Jews hesitate. Who wouldn't? So God famously says, "If you accept the Torah, good. And if not, this is where you'll be buried."[27] In a gallows-humor rap battle, the Lord is usually going to win.

It didn't take long for the Jews to get used to getting screwed. By the time they got to the land of milk and honey, I imagine that most of them were lactose intolerant (I know I am). Anyway, they didn't get to stay in the promised land very long before enemies were desecrating their temple and chasing them into exile yet again. Homelessness and marginalization are at the heart of the Jewish historical experience, which means that our humor is aimed at a generally hostile environment.

What's your image of "manna from heaven"? If you think of it as soft and melty and delicious, then you're a goy. For Jews, it's more like matzah in Purgatory—as in, "Oy, manna *again*?" It's the bread of aridity, the bread of never-ending affliction. The Jews got sick of manna pretty fast and started kvetching, waxing nostalgic about the melons they had enjoyed in Egypt—as slaves!—and about the ever-abundant fish, flavored with onions and garlic. And the Lord heard them and from on high said to Moses, "Fine, I'll send you some meat, and for the next month My Chosen People will eat nothing but quail, until it's coming out of your nostrils." And it did. And then the Lord sent a plague.[28]

It feels less innocent to enjoy the gallows humor of the Jewish Bible now that the Jewish State has become . . . less innocent. But if people are going to continue to invoke the Bible, we may as well emphasize its funnier and more universal elements. In the age of climate change, doesn't it seem at least a little ironic that virtually every culture on the planet tells stories about an ancient flood? Moreover, we've all known suffering; we've all felt desperate to lighten our despair with a joke. Spiritually, we've all walked through the desert or the valley of the shadow of death. So our current death march along what feels like the rapidly receding coastline of a small island nation is just a new variation on an old theme. At least we're all in it together, right?

OK, maybe not all of us. When Christian white supremacists marched in Charlottesville in the summer of 2017 and chanted, "You will not replace us! Jews will not replace us!" it was horrifying for both Jews and African Americans. And yet not really surprising. That's what happens to African Americans and Jews: we start to gain more leverage in society and then immediately become the scapegoats for all the idiotic mistakes made by the white Christian men who have been wielding power for the past several centuries.

The good thing about maintaining a really low baseline is that it's easier to be resilient. If you internalize the fact that the boulder is always going to roll back down the hill, then you can just roll *with* it. Try hitting a Jew while he's down: you'll miss every time, because that's precisely what he's expecting. Of course, you might occasionally surprise and confuse him with an act of kindness, but pretty soon he'll revert to his baseline.

An old Jewish man is sitting in a waiting room, sometimes rocking back and forth on his hands, sometimes leaning on a table. Every thirty seconds or so, he looks up to the heavens and says, "Oy, am I thirsty!" After a few minutes, the receptionist comes over and slams down a cup of water on the table. The old man is stunned for a moment, but then he drains the cup. Thirty seconds later, he looks up to the heavens: "Oy, was I thirsty!" Now that's an improvisation.

The old Jew knows from decades of experience that people are probably not going to be nice to him: either they'll ignore his thirst, or they'll quench it in an act of aggression, just to get him to shut up. So shutting up is not an option. By insisting on his never-ending misery, he's actually relieving it.[29]

Try asking an old Yiddish speaker, if you can find one, how things are going, and her answer is likely to be an Old World expression that means, essentially, "I feel like I'm lying in the earth baking bagels." *What?* Basically, "I'm as good as dead, but instead of resting peacefully, I'm slaving over a hot oven, spilling poppy seeds everywhere, making delicious baked goods *that no one is going to eat.*" For a Jewish mother, that is the definition of torture. Or Purgatory. But it's not Hell, because she still gets to tell you about it and make you feel like it's your fault. And then she goes back to rolling out the dough. Keep that manna coming![30]

Jews are good at Purgatory. In a way, Purgatory is the *perfect* baseline, because when truly horrific things happen, you don't have that far to go to make a full recovery. That's essentially how Jews approached life in many parts of Europe from the Middle Ages until the twentieth century. They were partially accepted in several different societies. They even thrived in many of them, living alongside Christians of various types—until suddenly something shifted, and it was time to round up the usual suspects again.

Q: What does "diaspora" mean?

A: It's a synonym for Purgatory.

The comedy of the shtetl could get quite dark. And as with a lot of dark African American comedy, shtetl jokes were typically meant for a members-only audience. Take the story of the shocking murder of a girl in a Ukrainian village. As soon as the news starts to spread, all the Jews flock to their run-down synagogue to consult: Should they flee? Should they take up arms? Should they agree on a communal alibi? No matter what, they realize, they will be blamed, and there will be a pogrom. But just then Mottel bursts through the back door. "I have wonderful news," he says, beaming. "The girl was Jewish!"[31]

Oy. To an unsympathetic ear, that's a horrific joke. The joke itself might spur a pogrom. It makes Jews seem utterly heartless. They're celebrating the death of a little girl in their community, somebody's adorable daughter? What's wrong with them? Let's burn down their synagogue and roast a pig in the embers.

In fact, most Jews are taught to uphold the sanctity of every individual life: that murder would of course have been considered a terrible tragedy. But conditions in Ukraine in, say, the eighteenth century, inspired a grim realism. If the girl had been Christian, then in the following days, dozens of Jews would have been killed. Far better for one family to be grieving than for the community to be decimated. It's a terrible calculation to have to make, but what really matters is the survival of the collective.

"The collective": that's rarely on the table anymore, though it did get invoked occasionally at the start of the COVID-19 pandemic. Couldn't we use a little dose of collectivist Jewish realism to help us with some of the choices and sacrifices we need to make to address climate change?

Western individualism is now so taken for granted that we find it almost unthinkable to sacrifice for the common good.[32] Every policy has to be framed as "win-win," which is code for "Sure, as long as it doesn't inconvenience me." Most millennials, who have grown up with the digital world at their fingertips, can't even handle minor annoyances, let alone real sacrifices. So why not call in some middle-aged American Jews? Hardship is like exercise for us: we know we need to practice coping, because things are only going to get worse. Selling a carbon tax to the American people would come naturally for a Jewish president. "Really, you can't handle a few extra pennies on your gas bill? You know, if Germany had passed a carbon tax in 1940, my Uncle Yitzhak would still be alive today."

Too soon? No, it's not. Hundreds of Holocaust survivors have gone on record saying that gallows humor helped create a sense of solidarity in the camps, which in turn helped countless people persevere through the seemingly endless horror. The only way to stay afloat amid widespread adversity is to recognize that everyone around you is in the same boat and that some of them are really funny. People not only told jokes in Auschwitz—they even organized secret cabarets and variety shows. At night, locked into cramped spaces, Jews would sing and dance and imitate the mannerisms of their guards and have contests to see who could compose the most creative insults in Yiddish. (This is the basic premise of Roberto Benigni's Oscar-winning film *La vita è bella* [*Life Is Beautiful*].) "We endured all of humanity's suffering," wrote one Holocaust survivor, "but we produced theater!" There's no business like Shoah business.[33]

The jokes that circulated among Jews in Europe in the late 1930s and early 1940s—in Germany and Holland, in the Warsaw and Lodz and Vilna ghettoes, in the concentration camps—ranged from joyous buffoonery to Nazi takedowns to the grimmest gallows humor. Like African slaves, Jews in World War II sometimes laughed to restore a basic sense of their humanity and their connection to others, and they sometimes laughed to resist the overwhelming power of their oppressors. Their engagement with comedy never erased the tragedy of their situation, but it occasionally helped them cope.

At the same time, of course, plenty of Jews avoided humor altogether, and plenty were too traumatized to respond to it, no matter how well intentioned and well delivered it might have been. Even twenty years later, when Mel Brooks made his hit movie *The Producers*, in which the characters stage a show called *Springtime for Hitler*, plenty of Jews objected that no one should ever make light of Nazi violence. And some people, today, reflexively object to

Moshe Pulaver performing in Auschwitz

climate-change jokes (if they ever hear any). But let them kvetch. Given how important a comedic perspective was for many people during the war—the title of one scholarly study of Holocaust humor is *It Kept Us Alive*—I think it's clear that we need a climate-change cabaret.

At my grandfather's funeral, when my cousins were asked to be pallbearers, one of them looked at the other and said, "Did Grandaddy change his name to Paul?" I was nineteen at the time and found the joke to be outlandishly inappropriate—and lame. All I wanted to do was cry. But now I see my cousin's question as a fairly brilliant improvisation, which helped steady him for a difficult job and which also honored our grandfather's addiction to punning—or paronomasia, as he called it, because that made it sound like a family illness, which it definitely was and still is. My grandfather was a Reform rabbi and biblical scholar who spent a great deal of time pondering the hereafter. He sometimes had sudden brainstorms that woke him up in the middle of the night. In a semitrance, he would find himself in the kitchen and ask himself, "Now what did I come here after?"

I know, not as funny as a good, solid Holocaust joke. Speaking of the hereafter, have you heard the one about the afterlife of Joseph Goebbels, Hitler's chief propagandist? God gives him a choice between Purgatory and Inferno. When Goebbels tours Purgatory, he finds it utterly boring—it's basically just a waiting room. But Hell has nightclubs with full bars and scantily clad dancers: Who cares if it's a little hot? Once he makes his decision, though, the bar scenes dissolve, and Goebbels sees that the reality of Hell is fiery pits and rivers of lava. Indignant, he turns to the Devil, and asks what it was he had witnessed on his tour. "Oh that?" the Devil says.

Circus poster from Theresienstadt, 1944

"That was just propaganda." (Feel free to replace "Goebbels" with "Trump" and "propaganda" with "fake news.")

How about the one in which Hitler goes down into the Underworld to speak with the spirit of Moses? The first thing Hitler says is, "Look, I'll let your people go if you tell me how you parted the Red Sea. I need to get across the English Channel so I can conquer England." Moses agrees to the deal and explains that all he had done was touch his magic rod to the waters. Hitler says, "Great! Where can I find that rod?" Moses says, "It's on display in the British Museum."

Q: What is fratricide?

A: When Hitler slaughters a pig.

Q: What is suicide?

A: If you ever tell this joke in Germany.

Two Jews are about to enter the gas chamber. One of them motions to the SS guard and says that he'd like to have a last glass of water before he dies. "Ssshh, Irving," the other Jew says—"don't make trouble!"

In the mid-1930s, at the time of the Dust Bowl in the western United States, there was also a drought in Germany. A rabbi in a small farming village came up with a radical idea: the whole community would pray for rain, and they would invite a local Nazi official to join them, since the Nazis were as desperate as the Jews for the drought to end. Of course, the congregants were scandalized. But the rabbi explained, "Look, if we make the Lord angry enough, maybe he'll send another flood."[34]

Please, let the waters rise! For Jews, as for African Americans, that's how bad it's gotten at times. Apocalyptic floods start to look like deliverance from oppression. So long as cultural memory persists, African Americans and Jews will be there to help us face dark realities. As an eventual follow-up to *Springtime for Hitler*, Mel Brooks developed a song-and-dance number about the Spanish Inquisition. And he collaborated with Richard Pryor to write the movie *Blazing Saddles* (1974), which got at the dark side of the mythical "winning" of the American West by exploring it from an unusual perspective: that of an African American sheriff (Brooks, meanwhile, plays a Native American leader who speaks Yiddish).[35]

It's certainly true that African Americans and Jewish Americans have had their misunderstandings over the years and that they continue to harbor a fair amount of mistrust for each other. But it's also worth noting that a shared sense of humor has sometimes brought the two communities together in powerful solidarity[36]—something I was reminded of while listening to Billy Crystal's eulogy for Muhammad Ali. It turns out that the two were friends for more than four decades, and it was comedy that sparked their connection. Ali, an outspoken advocate for civil rights and racial equality, was also the funniest pro athlete of the twentieth century, and a master of improv: "I'm so mean I make medicine sick!" "I should be a postage stamp—that's the only way I'll ever get licked!" When Crystal, a huge fan of such quips, started doing his Ali impersonation, he made fun of the boxer in a way that came across as an homage by an understudy, as if the two were sparring: "I'm so fast I can turn out the lights and be in bed before the room gets dark." Ali loved it. They weren't engaging in traditional gallows humor, but their jokes always had a clear political edge, given the historical context of their friendship and its symbolic significance. When Ali invited Crystal to go for runs with him at a country club near where they both lived, Crystal informed him that the club didn't allow Jews—and Ali never trained there again. Later in life, Ali acted as a co-sponsor for an initiative Crystal launched in Israel called Peace through the Performing Arts, which brings together Jews and Arabs to write and produce plays. Crystal explained that the project never would have had much credibility if not for the active, vocal support of the most famous Black Muslim in the world. Of course, in one version of Crystal's impersonation, Ali actually converts to Judaism: "I'm announcing tonight that I've got new religious beliefs. . . . I am now an Orthodox Jew, Izzy Izkowitz, *Chai*-am the Greatest of All Time!"[37]

We environmentalists are never going to be as funny as Billy Crystal and Muhammad Ali. But maybe we can help bring peace to the Middle East. In the Age of Climate Change, all of us are threatened by rising sea levels, which clearly makes Noah the most relevant prophet of the Five Books of Moses. Well, guess who also appears in the Quran?

Sura 10, verse 73: "We saved him and those with him in the ship, and made them viceroys (in the earth), while We drowned those who denied Our revelations. See then the nature of the consequence for those who had been warned."[38]

Turns out the Israelis and Palestinians have a lot more in common than just hummus and falafel: in the end, they're all climate refugees.

Q: Why did the Jews and Arabs wind up living in the desert?

A: It seemed like the safest place after the Flood.

3

INFERNO II

Even Hotter!

(Or, The Historic Depths and Rising Tide of Environmental Comedy)

Jean-Paul Sartre famously said that Hell is other people.[1] Certainly, that's been the experience of Jews and African Americans. But Hell is also the environment. Hell is all around you. You're never going to persevere without a sense of humor about nature.

Fortunately, for as long as there have been natural disasters, there have been natural-disaster jokes. Yet "environmental comedy," despite hints of a coming surge, is still almost as much of an oxymoron as "bipartisan cooperation." To my knowledge, no one has ever really studied or adopted the environmental comedy tradition—mostly because generations of ultraserious environmentalists have never had any interest in it, but maybe also because it trends very dark.

Perfect timing, no?

Dark genres have always come to the fore in dark times. Perhaps it makes good sense that the phrase "black humor" was first used by an antifascist, surrealist poet on the eve of World War II. For André Breton, dark comedy was *mostly* a modern phenomenon, honed to perfection by his contemporaries, artists like Picasso and writers like Kafka, who were attuned to the absurdities of mechanization and bureaucratization. But when Breton put together his *Anthology of Black Humor*, he traced the tradition all the way back to the early 1700s, and he argued that the origins of dark comedy lay in nature.

Breton's very first exemplary text is "A Modest Proposal," Jonathan Swift's satirical response to famine in Ireland. The essay's basic suggestion was that everyone might be better off if the poor sold their children to the wealthy—as snacks. "I have been assured," Swift wrote, ". . . that a young healthy child well nursed is at a year old a most delicious, nourishing, and wholesome food, whether stewed, baked, or boiled, and I make no doubt that it will equally serve in a fricassee, or a ragout." Now that's a win-win. For about a year, kids can survive almost entirely on mother's milk, but then they start to become a burden to their families, and if the families are not well off, then mom and dad and any siblings will all have to cut back on their consumption. So why not sell the kid, and actually boost your family's food-buying power?[2]

And in case you're not seeing the full environmental context here, we're talking about crop failures that were most likely caused by a climatic shift now known as the Little Ice Age. Global warming couldn't come fast enough for the Irish.[3]

Meanwhile, there's cannibalism. Actually, there's almost no socioenvironmental problem for which cannibalism isn't a solution. Couldn't we just eat all the billionaire oil executives? Or maybe, after following Bill Burr's advice and sinking some cruise ships, we could feed the bodies of the passengers to climate refugees. It would probably be convenient, since a lot of cruises stop at extremely impoverished port cities in hurricane zones, so that vacationers can ease their conscience by "supporting local businesses." This would just be a more direct kind of support.

The plight of climate refugees is no laughing matter, but I bet you'd find humor in the refugee camps, just like in the concentration camps. That's the thing about the horrors we inflict on each other and the horrors that nature inflicts on us to punish us for the horrors that we inflict on each other: for those who survive, life goes on, like it or not. So you'd better find a way to laugh.

People even chuckled sometimes about the ravages of the Black Death. One of Swift's contemporaries, Daniel Defoe (of *Robinson Crusoe* fame), wrote a novel about the 1665 plague that killed a quarter of London's population, and though much of the prose can only be described as lugubrious, there are many moments of comic relief—including lawyer jokes. Early in the book, our narrator prowls the London streets, finding a certain solace in the ghostly quiet: "The Inns-of-Court were all shut up. . . . Every Body was at peace, there was no Occasion for Lawyers." Always look on the bright side of death! And then come the doctor jokes: "Abundance of Quacks too died, who had the Folly to trust to their own Medicines." There's a useful lesson here. In a nutshell: you can't trust people trying to make a quick buck by offering to fix your problems. It's far better just to get used to having problems. Or, as one perfectly titled self-help manual from Defoe's era put it: *Preparation for Death the Best Preservative against Plague.*[4]

The disasters of early-modern Europe truly spurred some superlative dark comedy. Of course, they also spurred modernity, which many historians understand as society's fundamental effort to bring nature's chaos under control. Most Europeans thought of things like plagues as expressions of God's will, but modernizers thought of them as challenges to design better cities and implement new public-health measures. When a massive earthquake struck Lisbon in 1755, causing both fires and floods of biblical proportions, about 85 percent of the city's dwellings were destroyed and about fifteen thousand inhabitants perished. Enlightenment leaders immediately proposed that science and reason should be called on to reconstruct the metropolis with wider streets, a sewer system, and earthquake-resistant houses. Observing the horror from France, Voltaire wrote a poem urging compassion for the victims and endorsing the determined commitment to rebuild. But then he wrote his famous novel *Candide*, in which he ruthlessly mocked the people who tried to frame the earthquake as being "for the best," in "this best of all possible worlds," because, for instance, it provided so many new opportunities for expansive commercial development.[5]

Sure, the modern quest for security and stability has improved life for some people, but it's also just as delusive as the premodern goal of convincing God to stop showering us with catastrophes. "Now that Lisbon had been three-quarters destroyed by the earthquake," Voltaire wrote in *Candide*, "the local spiritual leaders could think of no more effective means of avoiding total ruin than providing the populace with a splendid *auto-da-fé*. The University of Coimbra had pronounced that the sight of a few people slowly burned to death with great ceremonial is the infallible recipe for preventing earthquakes."[6] Satirizing the human perpetrators of violence and cruelty might even help us accept the unavoidable violence of nature. Or, as Mel Brooks put it in his musical revue about the Spanish Inquisition: "Auto-da-fé? What's an auto-da-fé? It's what you oughtn't to do but you do anyway."[7]

Just over a century after the Lisbon quake, in the midst of the US Civil War, an obscure American humorist created a literary character called Petroleum Vesuvius Nasby, an ornery Northerner who sympathized with the South. Both of Nasby's given names signaled the natural explosiveness of his virulent racism. "Vesuvius" was probably a reference to a popular metaphor, coined by the Black abolitionist Frederick Douglass in 1849, which figured slavery as a "slumbering volcano"—an image meant to counter the slaveholders' position that the enslavement of Black people was the most natural thing in the world, since Africans were so passive and submissive.[8] And "Petroleum" was even more timely: it invoked the shocking discovery of oil in Pennsylvania in 1859, just two years before the Civil War started—an event that almost immediately transformed the US economy. (For starters, Pennsylvania crude allowed Nantucket and New Bedford to call back their whaling ships and put an end to the hunt for sperm oil.)[9]

GRAND BALL GIVEN BY THE WHALES IN HONOR OF THE DISCOVERY OF THE OIL WELLS IN PENNSYLVANIA.

Nasby's creator, David Ross Locke, published a long series of letters supposedly written by Nasby, much to the delight of President Lincoln, who sometimes read Nasby's adages out loud during Cabinet meetings, along with occasional "off-hand" puns about amputation (the newspapers published such jokes regularly). Yes, *that* President Lincoln, the one so often understood as depressive and almost always pictured as carrying the weight of the nation on his tall, skinny frame. Turns out he had a wicked sense of humor, which he thought of as a crucial coping mechanism: "If I did not laugh," he said, in response to the criticism he received for seeming to make light of apocalyptical events, "I should die." Petroleum V. Nasby helped him blow off steam by embodying the viciousness of the Democrats, Lincoln's most immediate political opponents, the people pushing hard to throw him out of office in the election of 1864 specifically because he had freed the slaves. In one memorable 1864 letter, Nasby described dressing up in blackface in Ohio and being horrified to find that some white farmers were so lacking in racism as to greet him warmly and offer him handsomely paid work.[10]

Like Voltaire, David Ross Locke and Abraham Lincoln recognized that no amount of modernization or Progress was going to end conflict, suffering, earthquakes, and volcanic eruptions. The Human Condition in Nature is inherently inflammable. In the moment, though, it can help to make jokes about both present and future conflagrations, whether caused by natural viciousness or cultural viciousness or (as is so often the case) some complex combination of the two.

The United States had its own memorable seismic disaster in 1906, when shifts along the San Andreas Fault set the city of San Francisco ablaze. What's not adequately remembered about the

1906 quake, though, is the strange atmosphere of lighthearted humor that it sparked. "All seemed to be merry," commented a San Francisco policeman. ". . . No matter where you went or who you spoke to, in the thick of that ruin with the fire blazing all around you, somebody found something to joke about." Three thousand people died, and about half of the city's survivors were suddenly homeless and without resources. Yet the record shows that this particular apocalypse was met with reserves of determined resilience and generous solidarity. There were bucket brigades, makeshift restaurants, improvised camps, faux hotels positioned over cracks in the earth so that they could offer free geothermal heating, and hand-drawn signs saying things like, "Eat, drink, and be merry, for tomorrow we may have to go to Oakland."[11]

Sociological studies show that calamities often call forth surprising amounts of altruism and mutual aid—something similar happened in the wake of Hurricane Katrina in 2005—but it certainly helps if you've already fostered a culture of good-humored adaptability. Though San Francisco was not free of the social divides that have always plagued US cities, by 1906 it already had some of the spirit of open-minded, feisty joyousness for which it is still known. "I can suggest no better 'cure,'" said the proud San Francisco writer Gertrude Atherton, just a month after the quake, "for those that live where nature has practically forgotten them and . . . in whom a narrow and prosperous life has bred pessimism and other forms of degeneracy, stunting the intelligence as well as atrophying the emotions, than to spend part of every year in earthquake country."[12]

Dark comedy thrived in the United States in the early twentieth century, even in the absence of further earthquakes, and the combination of war and depression ultimately produced one of the masterpieces of environmental satire in 1942: *The Skin of Our Teeth*, by Thornton Wilder, which was revived in 2017 in honor of climate change. It's a surrealist romp through tens of thousands of years of catastrophic human and natural history, set mostly in suburban New Jersey. In each of the three acts, the apocalypse seems imminent, but the main characters, the Antrobus family, survive each time, just barely, thanks to their desperate improvisations. (You could think of the play as a farcical work of Antrobology.) There's a fortune teller, a pet mammoth that needs to be milked every morning, and a beauty pageant, and at one point, Homer and Moses arrive at the Antrobuses' doorstep as climate refugees: apparently, a giant sheet of ice is rapidly approaching New Jersey from the north. But it soon becomes clear that the Ice Age is just one of many challenges this family has faced: "George," Mrs. Antrobus says, "remember all the other times. When the volcanoes came right up in the front yard. And the time the grasshoppers ate every single leaf and blade of grass, and all the grain and spinach you'd grown with your own hands. And the summer there were earthquakes every night." Plague, war, flood, famine, repeat.[13]

In the modern West, history is linear and progressive: each new era is perceived as utterly different from the last. Modernity destroyed our conception of cyclical time. But in *The Skin of Our Teeth*, it is always simultaneously the twentieth century and the Stone Age. George Antrobus, the hero of our story, is also Adam, the very first man ("he was once a gardener, but left that situation under circumstances that have been variously reported"). These days, he's always hard at work inventing and reinventing the wheel (literally: an oversized, primitive wheel is an important prop in the play). On the one hand, to survive the cold, he tells his family to "burn everything except Shakespeare"; on the other hand, he spends a day at the office trying to develop the first written language: "Why, when the alphabet's finished, we'll be able to tell the future and everything."[14]

The thing is, the future is always exactly the same: "I think I can prophesy . . . with complete lack of confidence," George says, "that a new day of security is about to dawn." In act 1, it's the Ice Age. In act 2, it's the deluge (and the Antrobuses escape on a small boat, with several pairs of animals). In act 3, it's a seven-year war fought over scarce resources, with Mr. Antrobus leading one side against a force commanded by his own son, Cain, whose name has been changed to Henry, in the hope that people will forget that he killed his brother, Abel. Throughout the play, Henry tries to wear his hair so that it covers up his telltale scar (the mark of Cain), but people always recognize him, and he keeps losing his temper and killing them with his slingshot. Will the violence ever end? No, probably not. But the Antrobuses always find a way to start over: "We've managed to survive for some time now, catch as catch can, the fat and the lean, and if the dinosaurs don't trample us to death, and if the grasshoppers don't eat up our garden, we'll all live to see better days, knock on wood." Near the end of the play, a character named Sabina (winner of the beauty pageant and George's sometime mistress) starts repeating lines that she first delivered in act 1. And then she turns to the audience: "This is where you came in. We have to go on for ages and ages yet. You go home. The end of this play isn't written yet. Mr. and Mrs. Antrobus! Their heads are full of plans and they're as confident as the first day they began,—and they told me to tell you: good night."[15]

Wilder is a realist about the power of nature and about humanity's failings—and he's a cynic about the "progress" of civilization—yet he leaves us smiling, because usually the future is far stranger than any apocalypse. At the start of *The Skin of Our Teeth*, the sun rises, and that causes a terrible shock for certain observers: "The Society for Affirming the End of the World went into a special session and postponed the arrival of that event for *twenty-four hours*."[16]

We environmentalists have been affirming the end of the world for some time now. But so far only a few pieces of the sky have fallen. Sure, things get crazy in Purgatory sometimes; sometimes it looks like we might slide down into Hell. We ought to have gotten better at protecting vulnerable people and ecosystems. Catastrophes do happen. But then we start over.

Wilder was writing about survival in the early years of World War II, just after André Breton compiled his highly inflammatory *Anthology of Black Humor*, which was at the printer in June 1940, when the Nazis were marching on Paris. Once the French capital fell, the puppet regime announced—within days—that all new books would have to be vetted by a censorship board. Breton's publisher did his best to apply pressure, but the censors took one look at the *Anthology* and started to laugh. Permission denied. The book didn't actually come out until after V Day.[17] By then, it almost seemed like old news, since the Good Guys had won and everyone was in a better mood. But it definitely came in handy as soon as the Cold War got rolling; all kinds of anxious artists embraced its sensibility.

I grew up in the culture of the Cold War, and I quickly got tired of watching made-for-TV movies about nuclear fallout and radiation sickness. Fortunately, I also had access to the darkly humorous albums put out by Tom Lehrer, a quirky, Jewish genius who never finished his PhD because he got too busy writing and performing satirical songs, at least one of which ("We Will All Go Together When We Go") made it seem like there might be something consoling, or even unifying, about Mutually Assured Destruction:

> When you attend a funeral,
> It is sad to think that sooner o'l-
> ater those you love will do the same for you.
> And you may have thought it tragic
> (Not to mention other adjec-
> tives) to think of all the weeping they will do.
> But don't you worry. . . .
>
> No more ashes, no more sackcloth,
> And an armband made of black cloth
> Will someday nevermore adorn a sleeve—
> For if the bomb that drops on you
> Gets your friends and neighbors too,
> There'll be nobody left behind to grieve!
>
> And we will all go together when we go—
> What a comforting fact that is to know.
> Universal bereavement—an inspiring achievement!
> Yes, we will all go together when we go!

Lehrer also wrote a song called "Pollution" in the early 1960s, echoing the message of *Silent Spring* but with a very different tone: "If you visit [an] American City, / You will find it very pretty. / Just two things of which you must beware: / Don't drink the water and don't breathe the air."[18]

The '60s also shaped Edward Abbey, one of the very few environmentalists who ever embraced comedy—partly because he was so annoyed by the seriousness and gentility of the nature-writing tradition. Abbey once threw a copy of *Walden* into an ammo can and dipped into it periodically while on a river-running expedition in Utah, which resulted in an uproarious essay-in-the-form-of-diary-entries called "Down the River with Henry Thoreau." Though Abbey acknowledged the considerable common ground he shared with his curmudgeonly forebear, he clearly chose to write the essay because of the opportunities it afforded him to compose creative takedowns. One morning, relishing some eggs, bacon, chilies, salsa, tortillas, fried potatoes, and his "usual breakfast beer," he found himself thinking that surely "Henry would not have approved of this gourmandizing. To hell with him. I do not approve of his fastidious puritanism. . . . Thoreau recommends a diet of raw fruits and vegetables . . . but confesses at one point to a sudden violent lust for wild woodchuck, devoured raw. No wonder; Henry was probably anemic."[19]

Abbey is not a writer for the twenty-first century. He too often comes across as a provocateur and a blowhard (not to mention a toxically masculine jerk). But there's quite a bit of fun to be had in considering his uncompromising commitment to wilderness and his downright Thoreauvian evisceration of industrial capitalism. His most famous novel, *The Monkey Wrench Gang*, which constructs a fantasy of blowing up the Glen Canyon Dam and letting the Colorado River run free again through the Wild West, has inspired many a radical, misanthropic tree hugger. And his most famous memoir, *Desert Solitaire*, is full of dark satire, as when Abbey imagines a Las Vegas–style animatronic Smokey the Bear monument at the entrance to every national park, rising against a billboard "gorgeously filigreed in brilliant neon and outlined with blinker lights, exploding stars, flashing prayer wheels, and great Byzantine phallic symbols that gush like geysers every thirty seconds." Press a button, and Smokey will tell you to "park your car, jeep, truck, tank, motorbike, snowmobile, jetboat, airboat, submarine, airplane, jetplane, helicopter, hovercraft, winged motorcycle, rocketship, or any other conceivable type of motorized vehicle in the world's biggest parkinglot behind the comfort station immediately to your rear." Other parts of the book read like scripts for late-'60s stand-up routines: "I prefer not to kill animals. I'm a humanist; I'd rather kill a *man* than a snake."[20]

For a more up-to-date brand of environmental humor, there's Michael Pollan, who, despite a fairly humane and genteel reputation, does kill animals. At least, he tries to. Especially when they threaten his garden. Pollan is best known as a food writer, but it turns out that he shares Abbey's obsession with Thoreau—and woodchucks. If Thoreau was too puritanical for Abbey, he is too laissez-faire for Pollan: enough of Henry's meditations on "how to *be* in nature," Pollan writes, in a book about gardening; what you really need to know, when a woodchuck comes for your vegetables, is "how to *act* there." And sometimes the action has to be decisive. After all, "this was about winning." So, he says, "I decided . . . to incinerate the woodchuck in his burrow." Thus begins a set piece that any environmentalist-in-training could study for a quick, hot lesson in self-directed humor:

> I had seen an item on the news concerning cabin fires aboard jetliners. In order to test a new, supposedly less combustible fuel, the FAA [Federal Aviation Administration] had simulated a cabin fire, and the footage they showed of fire racing wildly through the nar-

row enclosed space gave me an idea of exactly the sort of end the woodchuck deserved.

Take a moment to picture it.

So I poured maybe a gallon of gasoline down the burrow, waited a few minutes for it to fan out along the various passageways, and lit a match.

Evidently there was not much oxygen down there, because the flames shot in the wrong direction, up toward my face. I leapt back before I was singed too badly, and watched a black-orange fountain of flame flare out from the earth and reach for the overhanging olive bush. I managed to smother the fire with earth before the entire garden went up.

I guess this was my destroy-the-village-in-order-to-save-it phase. . . .

Fortunately, my brush with general conflagration among the vegetables shocked me out of my Vietnam approach to garden pests before I'd had a chance to defoliate my property or poison the ground water.[21]

I frequently go back to Pollan, and Abbey and Lehrer and Wilder and even Voltaire and Swift, to train myself in all the different kinds of dark comedy I've been touting in this book, from self-deprecation to improvisation to morale-boosting gallows humor to satirical admonishments. That's fine for professors like me and readers like you. (I love you for being a reader, but let's face it, we're probably both one-percenters when it comes to attention span.) What we really need, as we enter the third decade of the new millennium, is short video clips, memes, and elaborate publicity stunts, preferably involving celebrities. Fortunately, in the past few years, as more and more people have sought ways of grappling with climate change, the trickle of environmental comedy in Euro-American culture has started to look more like a steady stream.

It's never going to be easy to make climate jokes. All forms of dark comedy are risky and challenging, so it's not surprising that people have struggled to find the funny in hurricane trends and multivolume reports by atmospheric scientists. The British comedian Marcus Brigstocke referred to climate change as "far and away the most difficult comedy subject I've ever dealt with."[22] After all, it touches on spatial and temporal scales that we're just not wired to comprehend;[23] people laugh much more readily at the concrete realities that reach up and slap them in the face on a daily basis. Still, Brigstocke has stayed alert to opportunities: once he recited a Dr. Seuss–style poem about various shenanigans that occurred at some international climate negotiations (it was a great chance to do accents from around the world).[24] The implication was that climate negotiations might go a whole lot more smoothly if people approached them with a better sense of history (especially about colonialism and racism) and a better sense of humor.

Meanwhile, for those ordinary citizens just looking to be more responsible with regard to their everyday carbon emissions, my friend Jenny Price launched a fake advice column called "Green Me Up, JJ!" Example:

> Dear JJ:
> My 8-year-old son Rory wants to play in a baseball league, but the closest one is two towns away—28 miles! My wife wants to do it, but I think it's more important that Rory knows about climate change and learns how to act responsibly. Please advise!

JJ's answer suggests that the family could calculate its "family warming coefficient" by means of a "simple and very useful equation that two UC-Berkeley math whizzes have just developed." If the answer is less than one thousand, then they'd be good to go. If it's between one thousand and two thousand, then it would be OK to play only if the family converted its car engine to run on vegetable oil or if one of Rory's grandparents developed a life-threatening illness (thus reducing the family's overall carbon consumption). If the answer came out over three thousand, then Rory's participation in the baseball league would make him "personally responsible for a .004-inch rise in sea level." *Or* they could just find another family to carpool with.[25]

To cope with climate change, we need action at the individual level as well as the international negotiation tables; we need to overhaul local and state and national governments so that they're better able to improvise solutions; and we need to bring down fossil-fuel companies and replace them with businesses committed to clean, renewable, sustainable, efficient, job-rich energy models. Comedy is already supporting activism at all of these scales.

In fall 2017, a start-up based in Belgium announced excitedly that it would begin providing green electricity to Brussels by recruiting local refugees to ride bicycles that were wired into the grid—except the whole thing was a hoax conceived and carried out by an "artivist" group called We Are All Refugees, with help from the infamous comedic renegades known as the Yes Men. The project was "designed to call attention to the moral responsibility of developed countries for the so-called 'refugee crisis,' which is partly due to climate change (caused by rich countries) and partly due to colonial and current-day exploitation." There was even a perfectly vacuous promotional video.[26]

Two years before that, the Yes Men sent some actors to New York City to sell snow cones on the street using ice supposedly imported from the Arctic by Royal Dutch Shell, to promote the company's efforts to drill in regions that used to be inaccessible—before climate change. Well, guess what? Shell reacted to the publicity by abandoning its drilling plans. Sure, as the Yes Men themselves acknowledged, they were supported by the efforts of thousands of other antidrilling activists, but still: you should never underestimate the power of a satirical snow-cone cart.[27]

One of the Yes Men also delivered a commencement address a few years ago at his hippie-intellectual alma mater, Reed College, in Portland, Oregon. After some gentle jokes about twenty-first-century realities and some warm invitations to join The Revolution, the speaker, whose stage name is Igor Vamos, casually mentioned that over breakfast Reed's president had conveyed the news that the college was going to divest its endowment from the fossil-fuel industry. There was an immediate standing ovation. Vamos paused to look over his shoulder at the president, who was desperately trying to hide his shock. Yes, it was a prank. At the time (May 2014), colleges and universities were uniformly dismissive of the divestment movement: selling their fossil-fuel stock would be financially irresponsible; someone else would just buy it anyway; did students really want to give up all the benefits and conveniences of the modern energy economy? But Vamos never broke character or strayed from his script, which laid out Reed's (fake) plan to reinvest all of the college's fossil-fuel money in "community-owned renewable energy projects." Fast-forward a few years, and suddenly institutions of higher learning all over the world are announcing their divestment plans and their commitment to renewable energy. My university's board of trustees made the call in 2020, and Reed hopped on the bandwagon in the fall of 2021. It's hard to say no to the Yes Men.[28]

In any case, we should continue to make fun of fossil-fuel companies as ruthlessly as we can, since some people still believe that Shell and ExxonMobil and all the coal companies that no longer have the word "coal" in their names are actually serving society's best interests. In Australia, in 2014, a branch of the environmental group 350.org created a fake video advertisement for a fake coal company that was rolling out a new policy called "F— You!": "'F— You' means we can be passionate about our values, but not act on them," says the bespectacled, white, male president of the company. A sharply dressed blond woman, identified as the firm's chief financial officer, elaborates: "'F— You' takes what would be our present-day financial burden away from us and transforms it into a chronic, economic, social, cultural, and political crisis for future generations. . . . It ensures solid returns to our shareholders, by killing their grandchildren."[29] She could have added that widespread use of coal was already killing plenty of people and causing various crises and injustices in her own country, but she wasn't getting paid to care about the Wangan and Jagalingou natives marching in the outback to protest the leaching of coal slurry into their groundwater.[30]

The less ruthless, more laid-back brand of climate comedy has also been going strong recently, and it's especially helpful in bringing more people into the fold. Leading the way in encouraging this kind of "good-natured" humor are a group of mellow scholars at the University of Colorado (I hear it's very sunny there) who have been running a cutting-edge, web-based project called Inside the Greenhouse since 2012. Combining ecology, theater, geography, psychology, sociology, and communication, the programs sponsored by Inside the Greenhouse are meant to help students and activists find new ways of collaborating and reframing climate issues. Since 2016, there's been an annual Comedy and Climate Change Video Competition (with a three-minute limit for each entry) and an annual Stand Up for Climate Change event, at which students perform not just stand-up but also improv and sketch comedy. Yes, there have been some flops, but there have also been a lot more signs of hope and determination than you see at most universities.[31]

In 2020, the Stand Up show had to go online because of the COVID-19 pandemic, but it was full of bracing moments, like when a middle-school kid responds to a platitude about how young people are the solution to climate change by staring at the camera and saying, "I can't drive or vote, but it's up to me to save the f-ing planet?" The African American comedian Chuck Nice makes several appearances, interviewing a belligerent polar bear ("Where's *my* white privilege?"), doing a Black Trump skit, and urging us to "take our minds off the viral existential threat that faces all of humanity and focus on the *man-made* existential threat." The mood occasionally gets a little dark: at one point, Nice is barely able to speak the words "I can't breathe"—because his young daughter is choking him, out of frustration over his lousy homeschooling during the pandemic. Overall, though, the cuteness (and diversity) of the kids, the high spirits of the hosts (Beth Osnes and Max Boykoff, founders of Inside the Greenhouse), and the earnestness of the student-produced videos, combine to create an inclusive solidarity and even a glimmer of possibility. If we could sacrifice for the sake of those who are most vulnerable to COVID-19, then maybe we can also sacrifice for those who are most vulnerable to drought and pollution and rising sea levels.[32]

One of my favorite entries in the Colorado Video Competition was the very first winner, from 2016, titled "Weathergirl Goes Rogue," which was included as a special bonus in the 2020 online Stand Up show. Though Osnes and Boykoff and their colleagues tend to favor a more amicable brand of comedy (and they cite multiple studies to support that preference), they're at least open to gallows humor, and I think the cultural appetite for darker laughter is growing fast. That appetite is ultimately what pushed me to write this book. Good-natured jokes might be better for convincing people to listen to the science about climate change, but the massing hordes of depressives who have already listened now need a stronger brew—need direct acknowledgments of our current reality. The Rogue Weathergirl turns out to be a millennial forecaster (millennial in both senses), who, when asked to offer some details on the upcoming holiday weekend, instead goes on a reality-confronting rant: "This week," she explains, "satellites measured the smallest area of Arctic sea ice coverage in recorded history. That's four million square kilometers missing since the 1980s. . . . So instead of a bright white sheet reflecting sunlight back into space, that means dark water sucking up even more heat, making it melt faster and faster. Most scientists prefer the friendlier term of 'feedback loop,' but I'm more inclined to call it what it is: an inescapable death spiral."[33]

It's encouraging that millennials and Gen Zers are standing up and staging climate strikes and generally taking to the streets, with all their passion for justice and inclusion, and sometimes with the support of allied elders. Bridging the generation gap on climate issues is challenging, but gallows humor can help. In honor of Earth Day 2016, the folks at Funny or Die posted a one-minute video on their website titled "Old People Don't Care about Climate Change." Remarkably, it was originally produced by a forward-looking environmental organization dedicated to reaching people old enough to vote and young enough to care: Defend Our Future, an arm of the normally gloomy Environmental Defense Fund. The video features close-ups of elderly actors and celebrities who stare into the camera, looking grumpy and dismissive, and say, approximately, "Why should I spend my time worrying about climate change? I'll be dead by then." And just when you start wondering whether these selfish jerks care about anyone besides themselves—don't they at least want their grandchildren to have a future?—Ed Asner acknowledges that he does in fact think about posterity, just not in the way you might expect: "My grandkids are spoiled anyway," he says. "They could use a little hardship." Then Cloris Leachman gets even more sarcastic: "Ooooh, the big bad ocean's gonna rise up and swallow half of Florida!" At which point Bill Cobbs jumps in: "Good. That takes care of our country's Florida problem." The graphic at the end says that since the elderly don't give a hoot, you're going to have to—and the message works because it's not just another bromide about how kids are our future but rather a somewhat bitter acknowledgment of the fact that climate truly isn't a significant factor in the way that most older people vote in the United States.[34]

I also appreciate that Funny or Die directly encourages voting: you watch a video, and then get to judge whether it was funny or not. The stakes are high. Either enough of us say that the material is funny, or it gets deleted. Expunged. Eliminated. Wiped. Erased without a trace. Unless we take positive action, that carefully crafted one-minute recording will become an ex-video. It will no longer be taking up space on the internet.

Does this scenario remind you of anything? How about if I use words like "washed away" or "deluged" or "overwhelmed by the tides"? The stakes are high with climate change, too, so it's good that both environmentalists and comedians are starting to recognize the potential in dark environmental humor. That video about senior citizens has made the grade: it's still on the website as I write this, with eight million views and a rating of 84 percent funny.

Given the way Euro-American culture has been changing in the twenty-first century, young people are also primed to hear climate messages delivered in a queerer idiom. Nuclia Waste, a triple-nippled drag queen supposedly born of two former workers in the plutonium industry, might be the perfect spokesperson to help us accept the transnatural reality of Global Weirding. The environmental movement has long been in need of some alternative icons, not just to counter its congenital seriousness but also to unsettle its assumptions about what is normal, natural, and progressive. Far too often, climate activists (Al Gore leaps to mind) sound like they simply want to make America temperate again, replacing fossil fuels with scientifically proven renewables but maintaining absurdly high levels of energy consumption, so that certain kinds of families can continue having nice houses in the suburbs with two-car garages. A woke member of Gen Z, though, with their sex/gender/pronoun fluidity, will probably respond much more positively to someone like Nuclia Waste, who neither embraces nor rejects nuclear power but rather testifies to the truth that we are all complicit in the energy economy (which will never be perfectly clean) and all exposed to its fallout. Nuclia's three breasts, glittery beard, green hair, and neon party accessories, not to mention her perpetual feud with the purity-fetishizing "Miss Dee Contamination," who wears an "EPA-approved prom gown," serve to brace her audience for the unpredictable weather that's headed our way, and to promote an open-minded, adaptive, fabulous resilience.[35]

Something similar could be said about the *Climate Stew* blog and podcast, which, over the past few years, has been offering "a campy & thoughtful exploration of social justice, LGBTQ issues, and global warming." The activists behind *Climate Stew* mention sex whenever possible, make fun of polar bears (which lately seem to have been "hooking up" with grizzlies and "creating hybrid offspring known as prizzly bears or grolar bears"), and generally seek to remind people that the first step toward justice often involves accepting what seems terrifying (for instance, boundary-crossing sexual encounters). One of their key contributions has been to point toward the parallels between climate injustice and the HIV/AIDS pandemic of the 1980s: "You had a government that refused to acknowledge there was a problem, a president, Reagan, who wouldn't talk about it, a public that was hostile, afraid, and growing more hostile, and all the time people were suffering and dying." So why not imagine a climate-justice movement employing some of the strategies developed by successful anti-AIDS groups like ACT UP (the AIDS Coalition To Unleash Power)—"dramatic and playful public actions" relying on "giant puppets, elaborate costumes, and highly choreographed marches"?[36]

Some of the *Climate Stew* crew were previously members of Queers for the Climate, an activist collective known especially for its transgressive campaign called "It Gets Wetter." Playing on the idealistic "It Gets Better" campaign created by the gay sex guru Dan Savage, Queers for the Climate certainly didn't mean to undermine the encouragement Savage was offering to LGBTQ youth, but they did want to emphasize that there's nothing inevitable about progress. Sometimes, as Tig Notaro insisted when she did her comedy show about having cancer, things get worse, much worse, so we always need to be ready to come together, with a full activism toolkit, including a sense of humor, self-awareness, and, if possible, glee. Those who wanted to participate in the "It Gets Wetter" campaign were asked to "record a short video giving a reality check about climate change. Ideally, do this in drag, or scantily clad, or in a fabulous setting. The more outrageous the better."[37] The results, unsurprisingly, were outrageous. "I never imagined that I would have the right to marry the man I love," says Peterson Toscano in one of the clips. "I also didn't dream that it would happen just at the beginning of the end of the world. Yeah, thanks a lot, global warming!"[38] Righteous indignation can't hold a candle to campy indignation.

Things *have* gradually gotten better for some youth-oriented justice movements in the twenty-first century. As a college professor, I've been delighted to see more and more students out and proud, with regard to their sexuality, gender, race, ethnicity, and nerdy environmentalism (I'm thinking of you, Engineers for a Sustainable World). At the same time, though, I've seen a lot of young activists hamstrung by anxiety and depression, because some things have continued to get worse—as captured by a recent *New Yorker* cartoon in which a job interviewer asks, "And finally, where do you see yourself in five years?"—and the thought bubble of the interviewee shows him paddling a life raft through a midtown waterscape, in which only the top third of the Empire State Building is visible.[39]

Thankfully, young people are still watching Stephen Colbert, who has been doing gallows-humor bits about global warming for several years: "If you can't stop climate change, lay back and enjoy it. Now that it's totally futile, let's stop arguing about divisive issues like carbon taxes, alternative energy, or walking. Instead, we must come together and do what Americans have always done: we must burn hydrocarbons for pleasure. . . . I want to be clear: I'm not saying we just fiddle while Rome burns. I say we throw the fiddle on the fire too. I mean, that fiddle is good kindling, which we will need to keep the fire going as the floodwaters rise."[40] Yes, it's going to keep getting wetter.

On the last day of July 2019, just after a punishing thunderstorm, I attended an off-Broadway production called *Sea Level Rise: A Dystopian Comedy*. To be honest, it wasn't quite dystopian enough for me: we're not yet at the high-water mark for dark environmental humor. But I don't think this tide is going to be turned. The play is set in Florida, in 2048. Everything runs on Tidal Time, because at high tide, most of life becomes untenable in the town of Sweetwater. And now the hour of evacuation has arrived: Sweetwater's famous aquifer is about to turn into saline solution. That's the current dystopia.

But comedy is often about futurity. Classic comedies usually end with at least one wedding, suggesting a sense of potential: we'll make a new life together. Somewhere. And we do get hints of a love story in *Sea Level Rise*, as well as the story of a character who gets woke during the course of the play: Ana is a lawyer in her midthirties who has spent a decade working for a gigantic cruise line but is now ready to direct a climate-refugee relocation center. There's hope, in other words, despite the arrival of both Hell and high water. And the playwright, Henry Feldman, clearly wants you to relax into the reassuring feeling that you're watching an old-fashioned sitcom, with family members driving each other crazy, leaky septic tanks causing powerful odors, and a flirtatious sanitation inspector using technology to keep tabs on his favorite evacuee. Most of the jokes are of the appealing, good-natured variety; Feldman is forthcoming in interviews about how he consulted with Yale University's Program on Climate Change Communication, which takes the same basic approach as the Inside the Greenhouse project at the University of Colorado. Make people smile so that they're better able to assimilate the information you're trying to give them about global warming.

At the same time, though, Feldman allows some gallows humor into the play, through the desperate pronouncements of a climate scientist who teaches at the local college and whose didacticism is tempered by her alcoholism. Her wine- and tequila-fueled rants were my favorite parts of *Sea Level Rise*, and not just because I'm generally fond of drunken professors. She's like a pressure valve, allowing for the release of tension and frustration. "What should we do about climate change?" she asks at one point. And then she answers: "Whatever the heck we want. Because nothing we do is going to matter." Thank you for the affirmation![41]

Whether it's an inebriated physicist who says it or a Rogue Weathergirl or Stephen Colbert, a pronouncement of doom can make a huge difference. Sometimes you need to hear someone else say out loud what you're really feeling.

The irony is that, if they say it in the right tone of voice, you might find yourself not feeling it so intensely anymore.

EPILOGUE

Cold Comfort

Six common terms that no longer make sense in the age of climate change:

permafrost
one-hundred-year flood
heat wave
fire season
tropical disease
cold comfort[1]

"Cold comfort" dates to the fourteenth century—right at the dawn of the Little Ice Age, when suddenly, in northern Europe (not just in Ireland), summers stopped being dependably warm (and eventually crops started failing and people began to starve, etc., etc.). If you spoke some form of English, getting warmed up would have been synonymous with comfort, and it would have been rare. So the phrase "cold comfort" came to mean some small amount of superficial solace that ultimately just reminded you of how miserable you were.[2]

"Comfort" and "security" are the perfervid pipe dreams of ultramodernity. They seemed within reach in the twentieth century only because of our new supply of fossil fuels, which turned out to be not quite as limitless as we first thought. Moreover, a number of the "comforts" of modernity wound up creating additional insecurity: the people who gave you "better living through chemistry" also gave you toxic waste and cancer clusters. Plus, nuclear bombs. For almost all of history, most people have felt insecure and uncomfortable. It's my role, as a historian, to help you embrace that past. (Or you can just watch *Game of Thrones*, which conveys the same basic message.) The thing is, despite the occasional massive earthquake, insecurity and discomfort are not incompatible with basic contentment.

The British have known this for centuries. Long ago, they embraced the two most practical philosophies ever invented by Western culture: stoicism and epicureanism. If you're not familiar with them, they can be summed up as follows: keep calm and carry on; and enjoy that cup of tea while you still can. If you *are* familiar with them, then you probably also own a complete set of the Monty Python movies. (Given my stubbornly persistent concern with anachronism, I should note that tea drinking in the West did not come into the picture until the early seventeenth century, when European sailing ships started coming back from the East with all kinds of goodies. Of course, tea got even better shortly thereafter, when sugar imports spiked from slave colonies in the Caribbean. Basically, that epicurean cuppa is a pernicious symbol of imperial hubris and violence. But who am I to deprive the British of their last shred of comfort? You have to do whatever it takes to carry on. I'm addicted to coffee, and no, I am not willing to switch to locally sourced herbal concoctions.)

Speaking of the British, I've been worrying about them lately, given the precarious geography of their homeland. If not for Defoe, Swift, Edward Lear, Lewis Carroll, Douglas Adams, John Oliver, and Monty Python, I probably would have killed myself a long time ago. My wife and I even named one of our kids after Oscar Wilde. Which brings me to my Climate Change Limerick:

> The small island states are all skittish—
> 'Cause their coasts are as endangered as Yiddish.
> We've been anxious about the Maldivians
> And fretting about Papua New Guineans.
> But who's going to rescue the British?

I know, I know, not PC at all. In truth, I want to save *all* the small island nations. But I'd especially like to see how the British comedy tradition evolves in the age of global warming. Always look on the bright side of climate change!

Actually, there *is* a bright side. Siberian agriculture is looking better and better.[3] Also, people will hardly ever die of cold anymore, and cold has traditionally been deadlier than heat.[4] It's of course true that increased heat will cause deadly "tropical" diseases to spread into much higher latitudes, but some of us will survive, so the species should eventually develop resistance. Imagine: your great-great-granddaughter could be immune to Ebola and Zika! If she doesn't get killed by a coronavirus.

Regardless, what about the investment opportunities that climate change will bring? The future is so bright that you actually *will* have to wear shades—so why not design a new line of heavy-duty sunglasses? And every bed is going to need a new kind of "comforter." Swimming pools, carbon-free air conditioners, popsicles: there are endless opportunities for entrepreneurs, assuming we stick with a capitalist system. Or maybe climate change will shove us toward a new form of socialism? Which reminds me of a joke from eastern Europe:

> Q: What's the difference between capitalism and socialism?
> A: Capitalism is the exploitation of man by man, and socialism is the reverse.

And both are really good at exploiting nature. Actually, it's hard to imagine any human society that doesn't exploit nature to some extent. But not all exploitation is created equal. Who could object to exploiting solar energy? Global warming is going to make our transition to solar power that much simpler, because there will be more energy trapped within easy reach than ever before. Scolding environmentalists call it the Anthropocene, or the era of climate catastrophe; why not call it the golden age of solar energy?

Seriously, though: no more prophecies of doom, unless they're offered in a spirit of hijinks. Yes, we're dealing with some insanely complicated threats and challenges. But is climate change worse than the nuclear arms race? Is it worse than the Black Death? History is one long struggle to survive and thrive, and the main thing we need to do right now is renew the struggle. As a historian, I've come to believe that all we truly know about the future is that it will be different and surprising (though sometimes we do repeat certain kinds of mistakes, like war). The future will probably be bad for some people and good for others. Meanwhile, though, the present is providing us with plenty of evidence that the status quo isn't working for anyone beyond the white, uptight One Percent.

To revive our ungated communities, to embrace the collective good, to persist into the always-uncertain future, I think we need more hope and determination and less anxiety and fear. So a lot of my work as a historian has focused on digging up useful environmental ideas and practices that got buried by the backhoes and bulldozers of imperialism and industrial capitalism. Some of those ideas and practices even emphasized the joy to be found in nature, despite its overwhelming power and harsh limits. But, two books later, I realized that even joyful environmentalists tend to be congenitally serious, and not even the best environmental traditions will make a difference if all you know how to do is preach about them.

That's how I ultimately got to gallows humor. I feel like I finally found some genuine comfort in comedic traditions that, simply put, took the awe-inspiring power of nature a little less seriously. African Americans and Jews are really city people, after all. As Woody Allen admitted in *Annie Hall*, "I'm at *two* with nature."[5] Woodchucks are cute but also annoying. Maybe we environmentalists could do with a little less polar-bear-hugging and a little more environmental detachment. Maybe we could then accept the prospect of a slightly warmer future and refocus our attention on the people who are suffering today, right now. The planet is going to get through this, but the climate refugees may not, unless we start to care about them. (Look, I can't just make the preaching habit disappear, OK? I'm working on it.)

The simple reality is that the floodwaters are going to keep rising. Like they always have, to some extent. I think we'll be able to keep adapting—assuming we can get the young people to look up from their devices. In a way, it's a blessing that climate change has happened so quickly, because at least some of us have retained a few useful premodern skills, like house building and gardening. One more generation of sitting around surfing the internet in air-conditioned bliss, and we'd be toast. Or, in any case, we'd be even more obese than we already are, and lazier, and more entitled. Fortunately, in addition to our back-to-the-land millennials and Gen Zers who have spent time on the WWOOFing circuit (taking advantage of World Wide Opportunities on Organic Farms), we'll also be able to rely on many of the people from less industrial countries, since they've developed a basic resilience in the face of frequent blackouts and uncomfortably hot temperatures. People who experience adversity usually learn to improvise.[6]

Yes, it's true that we have our work cut out for us just to avoid Inferno. And it's true that Paradise is off the table. But we can do this. As Dick Gregory liked to say, "Hell hath no fury like a liberal scorned."[7] Environmentalists are finally starting to take themselves less seriously. And it also helps that we sometimes just stay at home eating hummus in our rock garden.

So: I'll see you in Purgatory. Even if we can never purge all our sins, it's better to make the effort than just to let ourselves be punished. Please, don't throw yourself in front of an SUV! We need you to help push the boulder and maybe teach some swimming lessons.

Q: How do you get to the Underworld?

A (Supplied by Aristophanes, in his play from 405 BCE, *The Frogs*): Well, the quickest ways are to hang yourself, drink poison, or throw yourself off a tower.[8]

Q: How do you get to Purgatory?

A: Relax—you're soaking in it.

But there's no guarantee that we get to stay here. So start collecting climate-change jokes. Tell them to your family and friends. Tell them to yourself when you're having a bad day. Shake your head and snort a little bit instead of slumping in your chair and sighing. Print some environmental cartoons in your local newsletter. The next time you give a lecture, make fun of how much of an earnest, righteous, humorless activist you used to be. Visit a school and do a Tom Lehrer sing-along (my thirteen-year-old son *loves* dark comedy, as long as it rhymes). Buy one of those hats or T-shirts that say, "Make Earth Cool Again." Talk to your local anarchist group about organizing a stealth operation to install solar panels at a coal-fired power plant.

I'm especially encouraged that younger people are starting to turn to gallows humor about climate change. (One youthful climate marcher recently carried a sign saying, "I was hoping for a cooler death.")[9] Young people have a lot to worry about these days. Several entirely serious magazine articles over the past few years have posited that climate change is causing millennials and Gen Zers to suffer from *pre*-traumatic stress disorder: they may seem fine right now, driving their Priuses down the leafy streets of their placid neighborhoods, but they're all too aware that the next hurricane or bomb cyclone could be just around the corner.

Pre-TSD was first reported, satirically, by *The Onion*, in 2006, in reference to soldiers training for deployment in Iraq (the relevant study was conducted by the Department of Future Veterans Affairs). But then climate change made it an actual medical syndrome. Yes, climate-induced pre-TSD is real.[10] Young people are just not going to make it without more comedy.

The even funnier thing is that these jokes have been told before. I don't know if the writers at *The Onion* knew about this history, but in 1936 tens of thousands of college students across the United States joined a satirical antimilitary organization called the Veterans of Future Wars (VFW). Some of the young women split off to form the Future War Spinsters, anticipating that they would never be able to marry because all the good men would be killed in their prime.[11]

It was a dark time in the United States. The Depression and Dust Bowl were in full swing. Millions of Great War veterans were still suffering from what was then known as shell shock. (That was a direct, honest way of talking about the long-term impact of war on soldiers; as George Carlin pointed out, though, in a 1990 stand-up routine about euphemisms, the terminology gradually shifted during the twentieth century: "shell shock" became "battle fatigue" in World War II and then "operational exhaustion" during the Korean War—"sounds like something that might happen to your car!"—and finally, in Vietnam, we got post-traumatic stress disorder.)[12] College students in the mid-1930s saw few prospects for themselves in the US economy, and they saw fascism on the rise in Europe. The founder of the VFW, Lewis Jefferson Gorin Jr., understood what most Americans still refuse to admit: that this nation is almost always at war.[13] Every few years, the US government is going to find another reason to sacrifice its young people. Gorin's generation was doomed.

Dark times demand dark comedy. Gorin knew that World War I veterans had spent years asking to collect bonuses from the federal government for their service overseas, so he figured that he and his peers ought to get an early start with their financial claims. "Common justice," he wrote, "demands that all of us who will be engaged in the coming war deserve, as is customary, an adjusted service compensation. . . . It is but common right that this bonus be paid now, for many will be killed or wounded in the next war." If you're dead, you can't collect a bonus. In April 1936, hundreds of students at Barnard College and Columbia University in New York City held a march featuring dolls dressed up as "future war orphans," with signs saying, "Spend your bonus here, not in the hereafter."[14]

No, the VFW did not end the Great Depression or prevent World War II (Gorin himself wound up fighting in Europe, just as he had expected to). And not even the best comedians or the most cutting-edge twenty-first-century satirical organizations will be able to prevent climate change. It's not about winning; there's never been a perfectly temperate climate anyway. It's about having the good humor and fortitude to continue to march for peace and justice and a relatively healthful environment. The key contribution of dark comedy, historically, is that it has helped people cope with dark realities. It discovers some levity in even the worst imaginable situations. It offers a shove in the direction of perseverance. It reminds us that we've been here before.

Climate trauma is real; it's not easy to face the floods and the chaotic weather and the unequal sacrifices. But at least Purgatorial struggle comes with a comforting familiarity.

ACKNOWLEDGMENTS

I've never liked the idea of "favorites." I'm too wishy-washy. I don't have a favorite food or a favorite book or a favorite landscape.

The one exception I'll confess to is that I have a favorite sound: my kids' laughter. Isn't that the whole point of dad jokes?

Sam, Abe, Ozzie: I love you. No joke. This book is for you, to carry with you into the future, as you work to make the world kinder and more joyous and more just. Feel the pain, but also keep laughing. And don't forget to look back into the past for inspiration.

Christine: thank you, again, always.

I felt my family's presence quite viscerally as I was writing this book. When I was cleaning out my parents' house in December 2015, after they had both died, I found a shelf of books I had never noticed before in my father's study: they were all about the history and theory of comedy. It felt like an invitation. And those books have been my constant companions ever since. I also frequently heard my mother's voice in my head, repeating my grandfather's jokes to my sister and me, making sure they got passed down from generation to generation. Look, on the road—a head!

My parents had a good sense of humor, but they were both quiet people. We smiled a lot when I was growing up; laughter was something we generally experienced in other houses. I learned to laugh mostly from my cousins, the Levines: Daniel, Jeremy, Sasha, and David, not to mention their parents, Uncle Joe and Aunt Bitsy. Uncle Joe officiated at my Jewish-Mormon wedding, and several of the guests asked me afterward if he was a rabbi or a stand-up comedian. Alas, I was a slow learner as a kid; my tendency was toward seriousness and sadness. But the Levines eventually taught me the kind of chill vigilance it takes to find humor in virtually any situation, and I will be forever grateful to them. Over the past decade or so, it has sometimes felt as though the Levines taught me the secret to life.

My sister, Debbie; her husband, Ricky; and their kids, Becca, Zach, and Nathaniel, continue to be a great source of fun and moral support, especially on Cape Cod, where I performed stand-up comedy for the first time a few years ago.

Yes, I started doing stand-up at open-mic nights while working on this project (though not in the small town where I live). I'm glad that life can be so surprising. Remembering how viscerally I hated any sort of spotlight when I was growing up, I'm still sort of befuddled that I felt called to the stage in this particular way. But it was thrilling as well as terrifying. On two occasions, once in Boston and once in New York, I invited a bunch of friends to come see me perform, and I will always be grateful for their warm support. Getting teased by old friends is one of life's great pleasures. Thank you to Midori Evans, Stephanie Koontz, Chris Malenfant, Naomi Meyer, Suzanne Mosher,

Charlie Radoslovich, Dan Stevens, and Naomi Tokisue in Boston; and to Vincent Balbarin, Daniel Evans, David Evans, Catherine Edwards, Jim Goodman, Ari Handel, Karen Harris, Brian Herrera, Mary Lui, Jenna Mammen, Lou Masur, Ilona Miko, and Sandy Zipp in New York. I wish I could see all of you much more often.

Thanks also to David Evans for recommending the *Good One* podcast (and thanks to Jesse David Fox for all those interviews with comedians).

Many friends and colleagues helped me with this project in innumerable ways, directly and indirectly. Eternal thanks to those who read the whole manuscript (in various forms): Charis Boke, Ben Cohen, Robert Johnston, Paul Lewis, Lou Masur, Tim Mennel, Cindy Ott, Charles Petersen, Jenny Price, Amy Reading, Mike Shohl, Carly Shonbrun-Siege, and Emily Turner. Carly also contributed some fantastic research assistance, and I hope she'll cling to the dream of combining comedy and environmentalism in her career.

Jenny, you're the best. Thanks especially for that ASEH variety show (and gratitude to Nicole Seymour and Barry Muchnick as well) and for all your pioneering work in environmental humor.

This book might never have made it without Brian Herrera. What an incredibly thoughtful, acute, giving friend you are—and hilarious, to boot. Thank you.

Brian shared my work with his coeditors Robin Bernstein and Stephanie Batiste, whose endorsement meant a huge amount to me and kept this project alive at NYU Press. Brian also urged me to meet with Sara Warner, and Sara wound up giving me just the pep talk I needed, though we barely knew each other. Sara, thank you for offering such warm and broad-minded encouragement.

Thanks to Cliff Kraft for hours of wide-ranging conversation and for *The Skin of Our Teeth*. Thanks to Joyce Chaplin for Jonathan Swift and for the endorsement of general hilarity. Thanks to Larry Moore for the insistence that lightness should always have a place in serious scholarship. Thanks to David Kay for a consistent spirit of engagement and for the Michael Wex book. Thanks to Nathan Connolly for bacon and donuts. Thanks to Mark Barrow for the James Watt cartoons. Thanks to Jeff Niederdeppe for all the work on climate communication and for taking the time to communicate with me. Thanks to Larry Buell for consistent support and enthusiasm and inspiring work over the past three decades, not to mention the tip about *The Comedy of Survival*. Thanks to David Outerbridge for coming to all those talks in metro Boston and for being a model of caring engagement for thirty-six years. Thanks to Sarah Rubenstein-Gillis for the steady supply of video clips. Thanks to John Forester for reading the first few pages I wrote on this material and for pushing me to keep going and figure it out. Thanks to John Demos for the most sustaining kind of mentorship.

Sometimes a few encouraging words can make a world of difference. Thanks so much to the friends and family members who kept me going on this project: Ernesto Bassi, Chris Bell, Rachel and Wayne Bezner-Kerr, Dorothea Braemer, Derek Chang, Fred Clarke, Sarah Ensor, Cristina Florea, Heather Furnas, Joe Giacomelli, Lou Greenberg, Rebecca and Reeve Hamilton, Bobbie Hauser, Josh Hauser, Becca Herson, Nick Howe, Sara Ivry, Julilly Kohler-Hausmann, Amy Kohout, Shiloh Krupar, Mary Lauppe, Carl Lee, Jessica Levman, Ben Liebman, Neil Maher, Laura Martin, Daegan Miller, Lynne Morrison, Nick Mulder, Victor Pickard, Sara Pritchard, Molly Reed, Dan Schmidt, Ruth Schmidt, Michael

Smith, Michael Trotti, Rob Vanderlan, Claudia Verhoeven, Josi Ward, and John Young.

Thanks to Cornell University—especially the History Department—especially Barb Donnell, Katie Kristof, Tamara Loos, Claire Perez, Georgie Saroka, Michael Williamson, and Judy Yonkin.

Several people at Cornell asked me to give presentations on this book's material, and those opportunities made a huge difference. Serious thanks, then, to Garrick Blalock, Peter Hess, Samsuda Khem-nguad, Johannes Lehmann, Scott MacDonald, and Susan Riha. Further afield, thanks so much to Michael Frederick and Tom Potter at the Thoreau Society, Bob Morrissey at the University of Illinois, and Bob Wilson at Syracuse University. And thanks to everyone who attended those performances, especially the people who followed up in various ways, by asking me questions, chatting with me afterward, sending me emails, or even, in the case of Patch Adams, sending me a handwritten letter. That feeling of connection and appreciation carried me a long way.

I'm deeply grateful to the people at NYU Press for believing in this book. Thanks especially to Eric Zinner and Furqan Sayeed and to the two anonymous reviewers, whose comments helped a great deal with my revisions. I appreciate the careful copyediting of Andrew Katz and production editing of Alexia Traganas. Thanks also to my agent, Zoë Pagnamenta, and to the fantastic permissions expert she recommended, Fred Courtright.

It was an intense pleasure to study comedic writing for a couple of years. I learned a huge amount (I hope) from Douglas Adams, Paul Beatty, Colin Beavan, Al Franken, David Gessner, David Grossman, Zarqa Nawaz, Michael Pollan, Mary Roach, David Sedaris, and Sarah Vowell. Also Shakespeare. Shakespeare is really, really funny.

And I owe a massive debt to the comedians who kept me going through the darkest days and years: Pamela Adlon, Mike Birbiglia, Lewis Black, Mel Brooks, George Carlin, Margaret Cho, Stephen Colbert, Billy Crystal, Tina Fey, Hannah Gadsby, Gary Gulman, Chelsea Handler, Moshe Kasher, Seth Meyers, Kumail Nanjiani, Trevor Noah, Tig Notaro, Emo Philips, Richard Pryor, Chris Rock, Sarah Silverman, Jon Stewart, Robin Williams, Ali Wong, Steven Wright, and the Yes Men.

Stay cool, everybody, and thanks for all the laughs.

ILLUSTRATION CREDITS

Purgatory cartoon: Megan Hills, "After waiting some time in Purgatory, Beryl decided to redecorate," © hills, www.mycartoonthing.com.

Al Gore: Kristin Callahan / Everett Collection / Alamy Stock Photo.

Bird hat: Public domain; digital image from Bygonely.

Polar bear cartoon: Courtesy of Liza Donnelly, https://lizadonnelly.com.

Phyllis Diller: Ronald Grant Archive / Alamy Stock Photo.

Dick Gregory: Pictorial Parade Archive / Getty.

Car-sharing poster: Government Printing Office for the Office of Price Administration/NARA Still Picture Branch (NWDNS-188-PP-42).

Moshe Pulaver: Public domain; digital image from Chaya Ostrower, *It Kept Us Alive: Humor in the Holocaust*, trans. Sandy Bloom (Jerusalem: Yad Vashem, 2014), 327.

Circus poster: Public domain; digital image from Chaya Ostrower, *It Kept Us Alive: Humor in the Holocaust*, trans. Sandy Bloom (Jerusalem: Yad Vashem, 2014), 301.

Grand Ball cartoon: Public domain; digital image from Alamy.

NOTES

PROLOGUE

1. Tig Notaro, *I'm Just a Person* (New York: Ecco, 2016), 139; and also see Notaro's long explanation for how she decided to begin her set with exactly this wording (129–39).
2. Gratitude to Brian Herrera for this question!
3. The show became a live album, whose title was *Live* (pronounced with a short *i*). You can listen to it on various platforms, for instance, Tig Notaro, "LIVE," YouTube, December 1, 2014, www.youtube.com/watch?v=oXk1DSbXsZk.
4. Among the many recent books on climate change that have contributed to my understanding are Kari Marie Norgaard, *Living in Denial: Climate Change, Emotions, and Everyday Life* (Cambridge, MA: MIT Press, 2011); Andrew J. Hoffman, *How Culture Shapes the Climate Change Debate* (Stanford, CA: Stanford University Press, 2015); Philip Smith and Nicolas Howe, *Climate Change as Social Drama: Global Warming in the Public Sphere* (New York: Cambridge University Press, 2015); Per Epsen Stoknes, *What We Think about When We Try Not to Think about Global Warming: Toward a New Psychology of Climate Action* (White River Junction, VT: Chelsea Green, 2015); Tracey Skillington, *Climate Justice and Human Rights* (New York: Palgrave Macmillan, 2017); and Maggie Nelson, *On Freedom: Four Songs of Care and Constraint* (Minneapolis, MN: Graywolf, 2021), 171–211.

5. A couple of relevant books that take a somewhat lighter approach are Rob Larson, *Bleakonomics: A Heartwarming Introduction to Financial Catastrophe, the Jobs Crisis, and Environmental Destruction* (London: Pluto Books, 2012), and Michael E. Mann and Tom Toles, *The Madhouse Effect: How Climate Change Denial Is Threatening Our Planet, Destroying Our Politics, and Driving Us Crazy* (New York: Columbia University Press, 2016). And kudos to Sam Nadell for his article "If Climate Change Is Real, Then Why Are There Oceans?," *McSweeney's*, July 5, 2016, www.mcsweeneys.net.
6. On the history and theory of comedy, see, for instance, Sigmund Freud, *Jokes and Their Relation to the Unconscious* (New York: Norton, 1989; orig. 1905); Paul Lauter, ed., *Theories of Comedy* (Garden City, NY: Anchor Books, 1964); Harry Levin, *Playboys and Killjoys: An Essay on the Theory and Practice of Comedy* (New York: Oxford University Press, 1987); Daniel Wickberg, *The Senses of Humor: Self and Laughter in Modern America* (Ithaca, NY: Cornell University Press, 1998); Ted Cohen, *Jokes: Philosophical Thoughts on Joking Matters* (Chicago: University of Chicago Press, 1999); Brian Boyd, "Laughter and Literature: A Play Theory of Humor," *Philosophy and Literature* 28 (2004): 1–22; Alenka Zupančič, *The Odd One In: On Comedy* (Cambridge, MA: MIT Press, 2008); Nichole Force, *Humor's Hidden Power: Weapon, Shield, and Psychological Salve* (Los Angeles: Braeden, 2011); Peter McGraw and Joel Warner, *The Humor Code: A Global Search for What Makes Things Funny* (New York: Simon and Schuster, 2014); Terry Eagleton, *Humour* (New Haven, CT: Yale University Press, 2019); Wayne Federman, *The History of Stand-Up: From Mark Twain to Dave Chappelle* (Beverly Hills, CA: Independent Artists Media, 2021); and David Steinberg, *Inside Comedy: The Soul, Wit, and Bite of Comedy and Comedians of the Last Five Decades* (New York: Knopf, 2021).
7. I'll generally use "dark comedy" in this book, to avoid confusion—since "black comedy" can refer either to dark comedy or to African American comedy. I'll capitalize the *B* when I'm using "Black comedy" to mean "African American comedy."

8. Steve Lipman, *Laughter in Hell: The Use of Humor during the Holocaust* (Northvale, NJ: Jason Aronson, 1991), 151. Also see Chaya Ostrower, *It Kept Us Alive: Humor in the Holocaust*, trans. Sandy Bloom (Jerusalem: Yad Vashem, 2014; orig. 2009); and David Slucki, Gabriel N. Finder, and Avinoam Patt, eds., *Laughter After: Humor and the Holocaust* (Detroit: Wayne State University Press, 2020).
9. Aaron Sachs, *Arcadian America: The Death and Life of an Environmental Tradition* (New Haven, CT: Yale University Press, 2013). Many environmental humanities scholars have been thinking along these lines in recent years; see, for instance, Anna Lowenhaupt Tsing, *The Mushroom at the End of the World: On the Possibility of Life in Capitalist Ruins* (Princeton, NJ: Princeton University Press, 2015); and Donna J. Haraway, *Staying with the Trouble: Making Kin in the Chthulucene* (Durham, NC: Duke University Press, 2016).
10. Herman Melville, *White-Jacket; or, The World in a Man-of-War* (1850; reprint, Evanston, IL: Northwestern University Press and the Newberry Library, 1970), 109. As it happened, my book about Melville and the trauma of modernity came out before this one: Aaron Sachs, *Up from the Depths: Herman Melville, Lewis Mumford, and Rediscovery in Dark Times* (Princeton, NJ: Princeton University Press, 2022).
11. Jenny Price, *Stop Saving the Planet! An Environmentalist Manifesto* (New York: Norton, 2021).
12. Gratitude also to Barry Muchnick, the fourth member of that sketch-comedy panel at the 2016 annual conference of the American Society for Environmental History in Seattle. See Nicole Seymour, *Bad Environmentalism: Irony and Irreverence in the Ecological Age* (Minneapolis: University of Minnesota Press, 2018). Though I situate my own work in history and American studies, it has been inspired in part by recent efforts within the environmental humanities to incorporate affect studies, performance studies, and queer studies, as embodied by Seymour's book. Other works along these lines that have been important to me include Catriona Mortimer-Sandilands and Bruce Erickson, eds., *Queer Ecologies: Sex, Nature, Politics, Desire* (Bloomington: Indiana University Press, 2010); Shiloh R. Krupar, *Hot*

Spotter's Report: Military Fables of Toxic Waste (Minneapolis: University of Minnesota Press, 2013); Richard D. Besel and Jnan A. Blau, eds., *Performance on Behalf of the Environment* (Lanham, MD: Lexington Books, 2014); Anthony Lioi, *Nerd Ecology: Defending the Earth with Unpopular Culture* (London: Bloomsbury Academic, 2016); and Kyle Bladow and Jennifer Ladino, eds., *Affective Ecocriticism: Emotion, Embodiment, Environment* (Lincoln: University of Nebraska Press, 2018). It's also worth noting that the book sometimes identified as the very first work of ecocriticism is Joseph W. Meeker, *The Comedy of Survival: Studies in Literary Ecology* (New York: Charles Scribner's Sons, 1974). Aspects of the book seem dated now, but it's nevertheless full of relevant gems: "Comedy demonstrates that man is durable even though he may be weak, stupid, and undignified" (24).

13. More and more scholars are seeking to connect psychology and political science; see, for instance, Gian Vittorio Caprara and Michele Vecchione, *Personalizing Politics and Realizing Democracy* (New York: Oxford University Press, 2017)—part of Oxford's Series in Political Psychology, edited by John T. Jost. In the humanities, scholars have been exploring the connection between the personal and the political in many different contexts but especially through affect studies; see, for instance, Ann Cvetkovich, *Depression: A Public Feeling* (Durham, NC: Duke University Press, 2012); and Brian Massumi, *Politics of Affect* (Cambridge, UK: Polity, 2015).
14. For a study of how the various arts have contributed to social movements, see T. V. Reed, *The Art of Protest: Culture and Activism from the Civil Rights Movement to the Present*, 2nd ed. (Minneapolis: University of Minnesota Press, 2019). Religion, music, and comedy were all incredibly important to African American resistance and resilience; see Lawrence W. Levine, *Black Culture and Black Consciousness: Afro-American Folk Thought from Slavery to Freedom* (1977; repr., New York: Oxford University Press, 2007). And Rebecca Solnit makes an elegantly broad argument about the significance of aesthetics to morale and politics, through a focus on gardening, in *Orwell's Roses* (New York: Viking, 2021).

15. Vine Deloria Jr., *Custer Died for Your Sins: An Indian Manifesto* (1969; reprint, Norman: University of Oklahoma Press, 1988), 147, 167. There's a whole chapter on "Indian humor" (146–67). Also see the recent book by Kliph Nesteroff, *We Had a Little Real Estate Problem: The Unheralded Story of Native Americans and Comedy* (New York: Simon and Schuster, 2021).
16. See Jason Guerrasio, "Leonardo DiCaprio Came Up with the Poignant Final Line in 'Don't Look Up,' Director Says," *Insider*, December 29, 2021, www.insider.com. *Don't Look Up* is surely a landmark in climate comedy, and I enjoyed every minute of it, especially when DiCaprio's character rants about denialism. But it often feels as though it's more about Trumpism and the COVID-19 pandemic than about climate change.
17. Albert Camus, *The Myth of Sisyphus and Other Essays*, trans. Justin O'Brien (New York: Vintage, 1955; orig. 1942), 88–92. Purgatory is actually a mountain in the *Divine Comedy*, as Joseph Meeker notes in *The Comedy of Survival*, and all the climbers know that "the purpose of climbing is to master oneself, not to conquer the mountain" (171). Indeed, I like to picture Sisyphus shaking his head and chuckling as he watches the boulder roll down the hill yet again. Comedy, as Meeker asserts, is "the art of accommodation and reconciliation" (38).
18. I'm paraphrasing. The translator Daniel De Leon renders the German this way: "Hegel says somewhere that all great historic facts and personages recur twice. He forgot to add: 'Once as tragedy, and again as farce.'" Karl Marx, *The Eighteenth Brumaire of Louis Bonaparte*, trans. Daniel De Leon (Ann Arbor, MI: C. H. Kerr, 1913), 9.
19. John Hersey, *Hiroshima* (New York: Vintage, 1989), 100.
20. A powerful, complicated, relevant study of the double edge of comedy is Paul Lewis, *Cracking Up: American Humor in a Time of Conflict* (Chicago: University of Chicago Press, 2006).
21. "Uproariously funny" is from Cassie da Costa, "The Funny, Furious Anti-Comedy of Hannah Gadsby," *New Yorker*, May 2, 2018, www.newyorker.com. The reviewer for the *Atlantic*, Sophie Gilbert, claimed that Gadsby simply "stopped being funny" about halfway into her set.

But the laughs kept coming—they were just more difficult laughs. See Gilbert, "*Nanette* Is a Radical, Transformative Work of Comedy," *Atlantic*, June 27, 2018, www.theatlantic.com. Also note Cynthia Willett and Julie Willett, *Uproarious: How Feminists and Other Subversive Comics Speak Truth* (Minneapolis: University of Minnesota Press, 2019), 150–53. Humanities professors all around the world have celebrated Gadsby's explicit use of her art history degree in the second half of her special to sharpen her barbs about the sexism of Western culture's worship of jerks like Picasso.

22. As Shiloh Krupar has put it, we environmentalists might have a lot to gain from engaging with "forms of collective play and artful endurance" (*Hot Spotter's Report*, 25). Also see Alison Bodkin, "Eco-Comedy Performance: An Alchemy of Environmentalism and Humor," in Besel and Blau, *Performance on Behalf of the Environment*, 51–72.
23. The most famous time was at the 1996 Democratic National Convention; see, for instance, Benny Johnson, "Al Gore Doing the Macarena," *BuzzFeedNews*, August 29, 2013, www.buzzfeednews.com.
24. See the clip from Gore's appearance on *The Late Show* on July 29, 2017: The Late Show with Stephen Colbert, "Get a Hot Date with Al Gore's Climate Change Pick-Up Lines," YouTube, July 29, 2017, www.youtube.com/watch?v=FCXxT94NJmA&list=LLT9YHSUqwruz7jPdQLdY6Hw&index=2294.
25. Quoted in Gerald Nachman, *Seriously Funny: The Rebel Comedians of the 1950s and 1960s* (New York: Pantheon Books, 2003), 227. Also see James Green, *Taking History to Heart: The Power of the Past in Building Social Movements* (Amherst: University of Massachusetts Press, 2000).
26. Nadia Y. Bashir, Penelope Lockwood, Alison L. Chasteen, Daniel Nadolny, and Indra Noyes, "The Ironic Impact of Activists: Negative Stereotypes Reduce Social Change Influence," *European Journal of Social Psychology* 43 (2013): 614. Of course, the stereotypes are not *entirely* the fault of the activists, but still: this is something we can work on.
27. The Onion, "New Prius Helps Environment by Killing Its Owner," YouTube, July 13, 2012, www.youtube.com/watch?v=bXEddCLW3SM.

28. It's worth noting that the phenomenon of gallows humor has been studied much less than one might think; I hope scholars in multiple disciplines will step up to the noose. One of the earliest studies dates back to World War II and explicitly emphasizes the way communities have used gallows humor to foster resilience under duress: Antonin J. Obrdlik, "'Gallows Humor': A Sociological Phenomenon," *American Journal of Sociology* 47 (March 1942): 709–16.
29. On the significance of joking among civil rights activists, see especially Stephen E. Kercher, *Revel with a Cause: Liberal Satire in Postwar America* (Chicago: University of Chicago Press, 2006), 284–85.
30. See, for instance, Roland Huntford, *Shackleton* (New York: Atheneum, 1986). Drag shows, sing-alongs, vaudeville acts, and comedy revues were common occurrences on polar expeditions. See Aaron Sachs, *The Humboldt Current: Nineteenth-Century Exploration and the Roots of American Environmentalism* (New York: Viking, 2006), 288–89. Note that the infamous Shackleton advertisement seems to have been a fake: "Men wanted for Hazardous Journey. Small wages, bitter cold, long months of complete darkness, constant danger, safe return doubtful" (quoted in Huntford, *Shackleton*, 365). No one has ever found any such ad. But it is pretty funny.
31. Ashley A. Anderson and Amy B. Becker, "Not Just Funny After All: Sarcasm as a Catalyst for Public Engagement with Climate Change," *Science Communication* 40 (August 2018): 524–40. But also see, especially, Christofer Skurka, Jeff Niederdeppe, Rainer Romer-Canyas, and David Acup, "Pathways of Influence in Emotional Appeals: Benefits and Tradeoffs of Using Fear or Humor to Promote Climate Change-Related Intentions and Risk Perceptions," *Journal of Communication* 68 (February 2018): 169–93. Special thanks to Jeff Niederdeppe for taking time to speak with me. I'm also grateful for the extensive discussion of the complex relationship between comedy and political activism in Sophia A. McClennen and Remy M. Maisel, *Is Satire Saving Our Nation? Mockery and American Politics* (New York: Palgrave Macmillan, 2014). And also note Paul R. Brewer and Jessica McKnight, "Climate as Comedy: The Effects of Satirical Television

News on Climate Change Perceptions," *Science Communication* 37 (October 2015): 635–57; Brewer and McKnight, "'A Statistically Representative Climate Change Debate': Satirical Television News, Scientific Consensus, and Public Perceptions of Global Warming," *Atlantic Journal of Communication* 25, no. 3 (2017): 166–80; and Guillaume Chapron, Harold Levrel, Yves Meinard, and Franck Courchamp, "Satire for Conservation in the 21st Century," *Trends in Ecology and Evolution* 33 (July 2018): 478–80. There's also a useful journalistic overview of this work: Elizabeth Preston, "Using Satire to Communicate Science," *Undark*, October 31, 2018, https://undark.org. It is certainly possible to see certain kinds of comedy, in certain contexts, as driving social and cultural change. See, for instance, Joanne R. Gilbert, *Performing Marginality: Humor, Gender, and Cultural Critique* (Detroit: Wayne State University Press, 2004); Rebecca Krefting, *All Joking Aside: American Humor and Its Discontents* (Baltimore: Johns Hopkins University Press, 2014); Matthew R. Meier and Casey R. Schmitt, *Standing Up, Speaking Out: Stand-Up Comedy and the Rhetoric of Social Change* (New York: Routledge, 2017); and Caty Borum Chattoo and Lauren Feldman, *A Comedian and an Activist Walk into a Bar: The Serious Role of Comedy in Social Justice* (Berkeley: University of California Press, 2020).

32. See, for instance, Julia Wilkins and Amy Janel Eisenbraun, "Humor Theories and the Physiological Benefits of Laughter," *Holistic Nursing Practice* 23 (November–December 2009): 349–54; and Ramon Mora-Ripoli, "The Therapeutic Value of Laughter in Medicine," *Alternative Therapies in Health and Medicine* 16 (November–December 2010): 56–64; and note the several scientific studies cited by the journalist Jordan Rosenfeld in his recent article, "11 Scientific Benefits of Having a Laugh," *Mental Floss*, April 11, 2018, http://mentalfloss.com. Once, when I gave a version of my "Climate Change Comedy Hour" lecture in Illinois, an elderly gentleman approached me afterward and asked me if I'd ever seen the movie *Patch Adams*, about a doctor, played by Robin Williams, who used comedy to help take care of his patients. I confirmed that I had, and he said, "Well, I'm Patch Adams." And he

was. And I love the work he does through his nonprofit organization, the Gesundheit! Institute. On framing climate change as a justice issue, see, for instance, Skillington, *Climate Justice and Human Rights*; Pope Francis's encyclical letter *Laudato Si: On Care for Our Common Home* (2015; available in many different editions); Wen Stephenson, *What We're Fighting for Now Is Each Other: Dispatches from the Front Lines of Climate Justice* (Boston: Beacon, 2016); Mary Robinson, *Climate Justice: Hope, Resilience, and the Fight for a Sustainable Future* (New York: Bloomsbury, 2018); Kum-Kum Bhavnani, John Foran, and Priya A. Kurian, eds., *Climate Futures: Re-imagining Global Climate Justice* (London: Zed Book, 2019); and Michael Méndez, *Climate Change from the Streets: How Conflict and Collaboration Strengthen the Environmental Justice Movement* (New Haven, CT: Yale University Press, 2020).

33. McClennen and Maisel insist that twenty-first-century satire in the US is constructive and activating specifically because it stands "in complete contrast to cynicism and other forms of negative humor" (*Is Satire Saving Our Nation?*, 17). I would suggest, though, that they may have skewed their argument by focusing almost exclusively on safe, mainstream, white, male comedians and that even those comedians come across as quite dark at times. In the end, I don't think it makes sense to claim that the impact of humor (in any form, anywhere along the "positive" to "negative" spectrum) will ever be predictable. The risk is part of what makes comedy fun—and powerful. Meanwhile, I am in complete agreement with the argument McClennen and Maisel make that in the twenty-first-century US, there is a lot of comedy explicitly designed to promote critical thinking and political engagement and that such comedy is sometimes demonstrably successful. I also share the cautious, open-ended hopefulness of Ted Gournelos and Viveca Greene, editors of the useful collection of essays *A Decade of Dark Humor: How Comedy, Irony, and Satire Shaped Post-9/11 America* (Jackson: University Press of Mississippi, 2011).
34. Two impressive scholars at the University of Colorado–Boulder have led this charge: Maxwell Boykoff in environmental studies and Beth

Osnes in theater and dance. They are among the directors of a cutting-edge project called Inside the Greenhouse: Re-telling Climate Change Stories: https://insidethegreenhouse.org. Also see their recent books: Boykoff, *Creative (Climate) Communications: Productive Pathways for Science, Policy, and Society* (New York: Cambridge University Press, 2019); and Osnes, *Performance for Resilience: Engaging Youth on Energy and Climate through Music, Movement, and Theatre* (New York: Palgrave Macmillan, 2017). Also note Beth Osnes, Maxwell Boykoff, and Patrick Chandler, "Good-Natured Comedy to Enrich Climate Communication," *Comedy Studies* 10 (June 2019): 224–36.

35. Michael Rothberg aptly describes the traumatized modern person as an "implicated subject," and in a very clear-headed and careful way, he opens up the question of whether trauma theory could be applied to climate change: see Rothberg, "Preface: Beyond Tancred and Clorinda—Trauma Studies for Implicated Subjects," in *The Future of Trauma Theory: Contemporary Literary and Cultural Criticism*, ed. Gert Buelens, Sam Durrant, and Robert Eaglestone (London: Routledge, 2014), xvi–xvii. Also see Rothberg, "Multidirectional Memory and the Implicated Subject: On Sebald and Kentridge," in *Performing Memory in Art and Popular Culture*, ed. Liedeke Plate and Anneke Smelik (New York: Routledge, 2013), 39–58.
36. See, for instance, Reed, *Art of Protest*, 63–64; Wickberg, *Senses of Humor*, 206; David Farber, *Chicago '68* (Chicago: University of Chicago Press, 1988), esp. 3–55; and Todd Gitlin, *The Sixties: Years of Hope, Days of Rage* (New York: Bantam, 1987), esp. 230–38, 320–36. "Mutual laughter tends to foster yet more mutuality," as Terry Eagleton has noted (*Humour*, 113). Eagleton also discusses dark comedy as seeking to "transcend the trauma in question without simply negating it, an exercise that demands both courage and truthfulness. As a way of liberating others into similar acts of confession, such dark humour is also a form of communication and comradeship" (141–42).

37. For some comedic analysis of these verses, see Hershey H. Friedman and Linda Weiser Friedman, *God Laughed: Sources of Jewish Humor* (New Brunswick, NJ: Transaction, 2014), 25, 65. On dark comedy as a source of solidarity, see especially Willett and Willett, *Uproarious*.

CHAPTER 1. INFERNO

1. See, for instance, Tim Flannery, *The Eternal Frontier: An Ecological History of North America and Its Peoples* (New York: Grove, 2001).
2. The indie actor was Jesse Eisenberg, in season 5, episode 12, "Under Pressure," which originally aired on January 15, 2014.
3. My immediate supervisor, John Young, was the savviest computer user on staff, so the password reflects his sense of humor. I actually used to "crunch numbers" for him, and I will always remember the time he bailed me out of a particularly difficult mathematical conundrum by looking at me with a twinkle in his eye and asking, "Have you ever read that book *How to Lie with Statistics*?" I had.
4. These are all real book titles. You could look them up.
5. All three of these quotations are from contemporary reviews quoted in M. Allen Cunningham, *Funny-Ass Thoreau* (Portland, OR: Atelier26, 2017), 19. Also note Laura Dassow Walls's eloquent comment, in her recent biography of Thoreau: "That he was . . . a lively and charismatic presence who filled the room, laughed and danced, sang and teased and wept, should not have to be said. But astonishingly, it does, for some deformation of sensibility has brought Thoreau down to us in ice, chilled into a misanthrope, prickly with spines, isolated as a hermit and a nag." I blame environmentalists. See Walls, *Henry David Thoreau: A Life* (Chicago: University of Chicago Press, 2017), xix.
6. Henry David Thoreau to Harrison Blake, May 20, 1860, in *Writings of Henry David Thoreau*, vol. 6, *Familiar Letters*, ed. F. B. Sanborn (Boston: Houghton Mifflin, 1906), 360.
7. Henry David Thoreau, *Walden* (Boston: Beacon, 2004), 69.

8. Henry David Thoreau, journal entry for January 3, 1861, in *The Journal of Henry D. Thoreau*, ed. Bradford Torrey and Francis H. Allen, vol. 14 (Boston: Houghton Mifflin, 1949), 306–7.
9. Thoreau, *Walden*, 67–68.
10. See Jennifer Price, *Flight Maps: Adventures with Nature in Modern America* (New York: Basic Books, 1999), 57–109.
11. On this period in environmental thought and activism, see, for instance, Robert Gottlieb, *Forcing the Spring: The Transformation of the American Environmental Movement*, rev. ed. (Washington, DC: Island, 2005; orig. 1993), 31–120; and Benjamin Heber Johnson, *Escaping the Dark, Gray City: Fear and Hope in Progressive-Era Conservation* (New Haven, CT: Yale University Press, 2017).
12. Both quoted in Johnson, *Escaping the Dark, Gray City*, 42, 38.
13. See Andrew C. Isenberg, *The Destruction of the Bison* (New York: Cambridge University Press, 2001).
14. See Price, *Flight Maps*, 1–55; quotation from Johnson, *Escaping the Dark, Gray City*, 21.
15. On the Progressive era, see, for instance, Michael McGerr, *A Fierce Discontent: The Rise and Fall of the Progressive Movement in America, 1870–1920* (New York: Oxford University Press, 2003); Jackson Lears, *Rebirth of a Nation: The Making of Modern America, 1877–1920* (New York: Harper, 2009); and Robert D. Johnston, "The Possibilities of Politics: Democracy in America, 1877 to 1917," in *American History Now*, ed. Eric Foner and Lisa McGirr (Philadelphia: Temple University Press, 2011), 96–124.
16. John Muir, *Our National Parks* (1901; repr., San Francisco: Sierra Club Books, 1991), 1–2.
17. Edgar Rice Burroughs, *Tarzan of the Apes* (1912; repr., New York: Ballantine Books, 1981), 110; pwgr2000, "The Tarzan Yell," YouTube, December 8, 2007, www.youtube.com/watch?v=1fQ631n50GI.
18. Muir, *Our National Parks*, 14, 21. Also see Mark David Spence, *Dispossessing the Wilderness: Indian Removal and the Making of the National Parks* (New York: Oxford University Press, 1999).
19. Muir, *Our National Parks*, 1, 3, 4, 14.

20. John Muir, "The Hetch-Hetchy Valley," *Sierra Club Bulletin* 6 (January 1908): 219–20; Phelan quoted in Thurman Wilkins, *John Muir: Apostle of Nature* (Norman: University of Oklahoma Press, 1995), 238.
21. See, for instance, Roderick Nash, *Wilderness and the American Mind*, 5th ed. (New Haven, CT: Yale University Press, 2014; orig. 1967).
22. John Muir, *My First Summer in the Sierra* (1911; repr., San Francisco: Sierra Club Books, 1988), 110. On connection and contamination, see Rebecca Solnit's powerful book *Savage Dreams: A Journey into the Landscape Wars of the American West* (1994; repr., Berkeley: University of California Press, 1999), esp. 144.
23. Rachel Carson, *Silent Spring* (1962; repr., Boston: Houghton Mifflin, 1994), 15, 32. For broader perspectives on this topic and this period of US environmental history, see, for example, Thomas R. Dunlap, *DDT: Scientists, Citizens, Public Policy* (Princeton, NJ: Princeton University Press, 1981); Samuel P. Hays, *Beauty, Health, and Permanence: Environmental Politics in the United States, 1955–1985* (New York: Cambridge University Press, 1987); Edmund Russell, *War and Nature: Fighting Humans and Insects with Chemicals from World War I to "Silent Spring"* (New York: Cambridge University Press, 2001); and Adam Rome, *The Genius of Earth Day: How a 1970 Teach-In Unexpectedly Made the First Green Generation* (New York: Hill and Wang, 2013).
24. The term exploded in popularity after Alvin Toffler developed the idea in his 1970 best-seller *Future Shock*, but it was already in use in the early to mid-'60s. See, for instance, Orrin E. Klapp, *Overload and Boredom: Essays on the Quality of Life in the Information Society* (Westport, CT: Greenwood, 1986), 5–9.
25. See, for instance, Kari Marie Norgard, *Living in Denial: Climate Change, Emotions, and Everyday Life* (Cambridge, MA: MIT Press, 2011).
26. C-SPAN, "Clip: President Obama's Anger Translator (C-SPAN)," YouTube, April 25, 2015, www.youtube.com/watch?v=HkAK9QRe4ds.
27. The quotations come from Carson, *Silent Spring*, 5, 10, 12, and 13.

28. See, for instance, Rebecca Krefting, *All Joking Aside: American Humor and Its Discontents* (Baltimore: Johns Hopkins University Press, 2014), 39–61; and Stephen E. Kercher, *Revel with a Cause: Liberal Satire in Postwar America* (Chicago: University of Chicago Press, 2006).
29. See especially Gerald Nachman, *Seriously Funny: The Rebel Comedians of the 1950s and 1960s* (New York: Pantheon Books, 2003); and Kercher, *Revel with a Cause*. Useful overviews of the history of comedy in the United States can be found in Krefting, *All Joking Aside*, 1–105; Russell Baker, ed., *Russell Baker's Book of American Humor* (New York: Norton, 1993); Kliph Nesteroff, *The Comedians: Drunks, Thieves, Scoundrels, and the History of American Comedy* (New York: Grove, 2015); and Sam Wasson, *Improv Nation: How We Made a Great American Art* (Boston: Houghton Mifflin Harcourt, 2017).
30. Jokes quoted in Nachman, *Seriously Funny*, 228, 224. Additional assertions from Diller: "First you do it to yourself, and then you have license to do it to others" (228); "To make it onstage, I had to make fun of myself first" (232); "Name me one comic who doesn't put himself down" (232).
31. Quoted in Kercher, *Revel with a Cause*, 410–11.
32. Quoted in Kercher, 404–6.
33. Quoted in Kercher, 415.
34. Quoted in Nachman, *Seriously Funny*, 404.
35. Quoted in Kercher, *Revel with a Cause*, 415.
36. In *Custer Died for Your Sins*, Vine Deloria Jr. argued that "Dick Gregory did much more than is believed when he introduced humor into the Civil Rights struggle. He enabled non-blacks to enter into the thought world of the black community and experience the hurt it suffered. When all people shared the humorous but ironic situation of the black, the urgency and morality of Civil Rights were communicated." Deloria, *Custer Died for Your Sins: An Indian Manifesto* (1969; repr., Norman: University of Oklahoma Press, 1988), 146. And see Kercher, *Revel with a Cause*, esp. 280–98; Krefting, *All Joking Aside*, 45–47; Nesteroff, *Comedians*, 214–31; and Mel Watkins, *On the Real Side: Laughing, Lying, and Signifying—The Underground Tradition of African-American*

Humor That Transformed American Culture, from Slavery to Richard Pryor (New York: Simon and Schuster, 1994), esp. 327–570. Also note William Schechter, *The History of Negro Humor in America* (New York: Fleet, 1970); Paul Beatty, ed., *Hokum: An Anthology of African-American Humor* (New York: Bloomsbury, 2006); Eddie Tafoya, *Icons of African American Comedy* (Santa Barbara, CA: Greenwood, 2011); Darryl Littleton, *Black Comedians on Black Comedy: How African-Americans Taught Us to Laugh* (New York: Applause Theatre and Cinema Books, 2016); and Danielle Fuentes Morgan, *Laughing to Keep from Dying: African American Satire in the Twenty-First Century* (Urbana: University of Illinois Press, 2020).

37. Quoted in Kercher, *Revel with a Cause*, 291.
38. Quoted in Nachman, *Seriously Funny*, 483, 486, 487.
39. Quoted in Nesteroff, *Comedians*, 220.
40. Quoted in Kercher, *Revel with a Cause*, 288.
41. Quoted in Nachman, *Seriously Funny*, 502–3.
42. See Kercher, *Revel with a Cause*, 286–98; Krefting, *All Joking Aside*, 45–46; Nesteroff, *Comedians*, 214–20; Littleton, *Black Comedians on Black Comedy*, 98–104; Watkins, *On the Real Side*, 495–503; and Tafoya, *Icons of African American Comedy*, 53–67.
43. P'Ville Pardner's Place, "Shortage of White People," YouTube, January 22, 2015, www.youtube.com/watch?v=ZxDP7lw7F8o. On Pryor, see Tafoya, *Icons of African American Comedy*, 35–72; Watkins, *On the Real Side*, 529–63; Littleton, *Black Comedians on Black Comedy*, 131–39; Richard Zoglin, *Comedy at the Edge: How Stand-Up in the 1970s Changed America* (New York: Bloomsbury, 2009), 41–64; and Scott Saul, *Becoming Richard Pryor* (New York: Harper Perennial, 2015).
44. Nina Shen Rastogi, "What's the Greenest Form of Birth Control?," *Slate*, March 3, 2009, https://slate.com.
45. Team Coco, "Bill Burr's Solution to Environmental Problems | Conan on TBS," YouTube, January 3, 2017, www.youtube.com/watch?v=WKbBDKsSEic.
46. On the question of population growth and its relationship to environmentalism in this period, see, for instance, Thomas Robertson, *The*

Malthusian Moment: Global Population Growth and the Birth of American Environmentalism (New Brunswick, NJ: Rutgers University Press, 2012); and Paul Sabin, *The Bet: Paul Ehrlich, Julian Simon, and Our Gamble over Earth's Future* (New Haven, CT: Yale University Press, 2013).

47. Dr. Seuss, *The Lorax* (New York: Random House, 1971).
48. See, for example, Gottlieb, *Forcing the Spring*, esp. 1–29, 161–409; and Mark Dowie, *Losing Ground: American Environmentalism at the Close of the Twentieth Century* (Cambridge, MA: MIT Press, 1995).
49. amiableamy24, "The West Wing-Pluie," YouTube, March 13, 2009, www.youtube.com/watch?v=Avoo-8GvBlA.
50. The Onion, "Tips for Combating Climate Change," Twitter, July 25, 2017, http://twitter.com/theonion/status/890050311196254211.
51. The book is also quite funny; its subtitle is "The Adventures of a Guilty Liberal Who Attempts to Save the Planet and the Discoveries He Makes about Himself and Our Way of Life in the Process." Colin Beavan, *No Impact Man* (New York: Farrar, Straus and Giroux, 2009).

CHAPTER 2. PURGATORY

1. Oscars, "Chris Rock's Opening Monologue," YouTube, March 23, 2016, www.youtube.com/watch?v=kqhVNZgZGqQ.
2. See, for instance, Niraj Chokshi, "How #BlackLivesMatter Came to Define a Movement," *New York Times*, August 22, 2016, www.nytimes.com.
3. Tim Baysinger, "This Is How Many People Watched the Oscars This Year," *Time*, February 27, 2017, http://time.com (the article also cites data for 2016).
4. See Stephen Nachmanovitch, *Free Play: Improvisation in Life and Art* (New York: Tarcher/Putnam, 1990); Kelly Leonard and Tom Yorton, *Yes, And: How Improvisation Reverses "No, But" Thinking and Improves Creativity and Collaboration* (New York: Harper Business, 2015); Alan Alda, *If I Understood You, Would I Have This Look on My Face? My Adventures in the Art and Science of Relating and Communicating* (New York: Random House, 2017); and Sam Wasson, *Improv Nation: How We*

Made a Great American Art (Boston: Houghton Mifflin Harcourt, 2017).

5. I'm convinced of the political possibilities of gallows humor, but one should also be aware of its limitations: see Paul Lewis, "Three Jews and a Blindfold: The Politics of Gallows Humor," in *Semites and Stereotypes: Characteristics of Jewish Humor*, ed. Avner Ziv and Anat Zajdman (Westport, CT: Greenwood, 1993), 47–57.
6. Melonhead622, "Eric Idle—'Always Look on the Bright Side of Life'—Stereo HQ," YouTube, February 11, 2011, www.youtube.com/watch?v=SJUhlRoBL8M.
7. This is one of the fundamental arguments of my book *Arcadian America: The Death and Life of an Environmental Tradition* (New Haven, CT: Yale University Press, 2013).
8. Antonin J. Obrdlik, "'Gallows Humor': A Sociological Phenomenon," *American Journal of Sociology* 47 (March 1942): 709. If one reads across the comedy traditions of historically oppressed peoples, the connection between gallows humor and satire becomes fairly clear. See, for instance, Rebecca Krefting's examination of what she calls "charged humor" in *All Joking Aside: American Humor and Its Discontents* (Baltimore: Johns Hopkins University Press, 2014); as well as Joanne R. Gilbert, *Performing Marginality: Humor, Gender, and Cultural Critique* (Detroit: Wayne State University Press, 2004); and virtually any of the works I cite in other notes on African American or Jewish humor. For some deeper perspectives on dark/black comedy, see Bruce Jay Friedman, ed., *Black Humor* (New York: Bantam, 1965); U. C. Knoepflmacher, *Laughter and Despair: Readings in Ten Novels of the Victorian Era* (Berkeley: University of California Press, 1973); Patrick O'Neill, *The Comedy of Entropy: Humour, Narrative, Reading* (Toronto: University of Toronto Press, 1990); Alan R. Pratt, ed., *Black Humor: Critical Essays* (New York: Garland, 1993); and Laurie Stone, *Laughing in the Dark: A Decade of Subversive Comedy* (Hopewell, NJ: Ecco, 1997).
9. The Mets, amazingly, went from worst to first in only seven years, which is the kind of fact that has made sports fans particularly good

at retaining hope in dark times. As for the scholarly question of periodizing dark/black comedy, see especially Friedman, *Black Humor*; Knoepflmacher, *Laughter and Despair*; and Pratt, *Black Humor.*

10. Quoted in William Schechter, *The History of Negro Humor in America* (New York: Fleet, 1970), 11. Also see Mel Watkins, *On the Real Side: Laughing, Lying, and Signifying—The Underground Tradition of African-American Humor That Transformed American Culture, from Slavery to Richard Pryor* (New York: Simon and Schuster, 1994); Paul Beatty, ed., *Hokum: An Anthology of African-American Humor* (New York: Bloomsbury, 2006); Eddie Tafoya, *Icons of African American Comedy* (Santa Barbara, CA: Greenwood, 2011); and Darryl Littleton, *Black Comedians on Black Comedy: How African-Americans Taught Us to Laugh* (New York: Applause Theatre and Cinema Books, 2016).
11. James Baldwin articulated this idea in more general terms in 1955: "It began to seem that one would have to hold in the mind forever two ideas which seemed to be in opposition. The first idea was acceptance, the acceptance, totally without rancor, of life as it is, and men as they are: in the light of this idea, it goes without saying that injustice is a commonplace. But this did not mean that one could be complacent, for the second idea was of equal power: that one must never, in one's own life, accept these injustices as commonplace but must fight them with all one's strength." See Baldwin, *Notes of a Native Son* (1955; repr., Boston: Beacon, 2012), 114–15.
12. See, for instance, Andrew Nikiforuk, *The Energy of Slaves: Oil and the New Servitude* (Vancouver: Greystone Books, 2012).
13. Quoted in Watkins, *On the Real Side*, 39–40.
14. See, for instance, Philip Bump, "Jim Inhofe's Snowball Has Disproven Climate Change Once and for All," *Washington Post*, February 26, 2015, www.washingtonpost.com.
15. Watkins, *On the Real Side*, 448, 132.
16. See, for instance, Lewis Hyde, *Trickster Makes This World: Mischief, Myth, and Art* (1998; repr., New York: North Point, 1999), esp. 110–35.

17. Carter's speech actually received some praise at first, but by the presidential campaign of 1980, it had become an albatross. See, for instance, Kevin Mattson, *"What the Heck Are You Up To, Mr. President?": Jimmy Carter, America's "Malaise," and the Speech That Should Have Changed the Country* (New York: Bloomsbury, 2009); and NPR, "Examining Carter's 'Malaise Speech,' 30 Years Later," July 12, 2009, www.npr.org.
18. Peter McGraw and Joel Warner, *The Humor Code: A Global Search for What Makes Things Funny* (New York: Simon and Schuster, 2014), 23.
19. reece888888, "Jimmy Carter's Full 'Crisis of Confidence' Speech (July 15, 1979)," YouTube, May 2, 2012, www.youtube.com/watch?v=kakFDUeoJKM.
20. evelkidnievel, "George Carlin—Americans, Shopping & Eating," YouTube, November 28, 2013, www.youtube.com/watch?v=YQ8jp88_O4g.
21. The poster is reprinted in Connie Chiang, "Winning the War at Manzanar: Environmental Patriotism and the Japanese American Incarceration," in *Rendering Nature: Animals, Bodies, Places, Politics*, ed. Marguerite S. Shaffer and Phoebe S. K. Young (Philadelphia: University of Pennsylvania Press, 2015), 238.
22. Watkins, *On the Real Side*, 469–71; Hyde, *Trickster Makes This World*, 271–73.
23. See Hyde, *Trickster Makes This World*, esp. 20–21; and also Rebecca Solnit, *Hope in the Dark: Untold Histories, Wild Possibilities* (New York: Nation Books, 2006), 77–80.
24. NPR, "A Less Restrained Obama Finally Says 'Bucket,'" June 29, 2015, www.npr.org; for the bacon and donuts clip, see MeLoad Virals, "President Obama Destroys Republicans over Climate Change!!," YouTube, November 5, 2015, www.youtube.com/watch?v=7Xkpho33WHM.
25. See, for instance, Henry D. Spalding, ed., *Encyclopedia of Jewish Humor* (New York: Jonathan David, 1969); Chaim Bermant, *What's the Joke? A Study of Jewish Humor through the Ages* (London: Weidenfeld and Nicolson, 1986); Hershey H. Friedman and Linda

Weiser Friedman, *God Laughed: Sources of Jewish Humor* (New Brunswick, NJ: Transaction, 2014); Michael Krasny, *Let There Be Laughter: A Treasury of Great Jewish Humor and What It All Means* (New York: William Morrow, 2016); and Jeremy Dauber, *Jewish Comedy: A Serious History* (New York: Norton, 2017).

26. See especially Michael Wex, *Born to Kvetch: Yiddish Language and Culture in All of Its Moods* (New York: Harper Perennial, 2006).
27. This line is from the Talmud; it's a standard interpretation of Exodus 19 and the following chapters. See Wex, 8; and also note Friedman and Friedman, *God Laughed*, 39–45.
28. See Numbers 11; Bermant, *What's the Joke?*, 11–12; and Friedman and Friedman, *God Laughed*, 61.
29. Michael Wex uses a version of this joke to open his book *Born to Kvetch* (1–3); and it appears in virtually every other book about Jewish humor.
30. Wex, 40–42.
31. Friedman and Friedman, *God Laughed*, 5–6.
32. For a couple of recent reflections on this issue, see Robert B. Reich, *The Common Good* (New York: Random House, 2018); and Jedediah Purdy, *This Land Is Our Land: The Struggle for a New Commonwealth* (Princeton, NJ: Princeton University Press, 2019).
33. See Spalding, *Encyclopedia of Jewish Humor*, 184–200; Alan Dundes and Thomas Hauschild, "Auschwitz Jokes," *Western Folklore* 42 (October 1983): 249–60; Steve Lipman, *Laughter in Hell: The Use of Humor during the Holocaust* (Northvale, NJ: Jason Aronson, 1991), esp. 152–54; Chaya Ostrower, *It Kept Us Alive: Humor in the Holocaust*, trans. Sandy Bloom (Jerusalem: Yad Vashem, 2014; orig. 2009), esp. 229–330 (quotation on 330); and David Slucki, Gabriel N. Finder, and Avinoam Patt, eds., *Laughter After: Humor and the Holocaust* (Detroit: Wayne State University Press, 2020).
34. Lipman, *Laughter in Hell*, 50, 83, 52, 193, 84.
35. The Spanish Inquisition number appears in Brooks's film *The History of the World, Part I* (1981). The sheriff's role in *Blazing Saddles* was

meant for Pryor, but in the end it went to Cleavon Little. On the intriguing links between Native American and Jewish American culture, see Alan Trachtenberg, *Shades of Hiawatha: Staging Indians, Making Americans, 1880–1930* (New York: Hill and Wang, 2004), esp. 140–69.

36. Note that Lawrence Levine, celebrated Jewish historian of African American culture, has a powerful chapter called "Black Laughter" in his classic book *Black Culture and Black Consciousness: Afro-American Folk Thought from Slavery to Freedom* (1977; repr, New York: Oxford University Press, 2007), 298–366; the parallel between African American humor and Jewish humor becomes explicit on 335–36.
37. ABC News, "Muhammad Ali Funeral | Billy Crystal Imitates Ali," YouTube, June 10, 2016, www.youtube.com/watch?v=7XB3sD9QJCI; "30 of Muhammad Ali's Best Quotes," *USA Today*, June 5, 2016, www.usatoday.com. Also see, for instance, Mike Marqusee, *Redemption Song: Muhammad Ali and the Spirit of the Sixties* (New York: Verso, 1999); and Jonathan Eig, *Ali: A Life* (Boston: Houghton Mifflin Harcourt, 2017).
38. I used the English translation by Muhammad M. Pickthall, available at Gregory R. Crane, ed., Perseus Digital Library, Tufts University, www.perseus.tufts.edu.

CHAPTER 3. INFERNO II

1. "L'Enfer, c'est les autres"—from Sartre's 1945 play *Huis Clos*, translated as *No Exit*; see Jean-Paul Sartre, *"No Exit" and Three Other Plays*, trans. Stuart Gilbert (New York: Vintage, 1955; orig. 1945), 47.
2. André Breton, *Anthology of Black Humor*, trans. Mark Polizzotti (San Francisco: City Lights Books, 1997; orig. 1945); Swift quotation on 11.
3. See, for instance, Brian Fagan, *The Little Ice Age: How Climate Made History, 1300–1850* (New York: Basic Books, 2001); and Philipp Blom, *Nature's Mutiny: How the Little Ice Age of the Long Seventeenth Century Transformed the West and Shaped the Present* (New York: Liveright, 2019).

4. Daniel Defoe, *A Journal of the Plague Year*, ed. Louis Landa, introd. David Roberts (New York: Oxford University Press, 2010; orig. 1722), 16, 32, xi.
5. François-Marie Arouet de Voltaire, *Candide, or Optimism*, trans. Joan Spencer, in *"Candide" and Other Stories* (London: Oxford University Press, 1966; orig. 1759), 128, 116.
6. Voltaire, 129.
7. *History of the World, Part I*, dir. Mel Brooks (20th Century Fox, 1981).
8. Herman Melville made use of the same phrase in his story "Benito Cereno"; see Robert K. Wallace, *Douglass and Melville: Anchored Together in Neighborly Style* (New Bedford, MA: Spinner, 2005), 101–2, 110–17.
9. See, for instance, Brian Black, *Petrolia: The Landscape of America's First Oil Boom* (Baltimore: Johns Hopkins University Press, 2000).
10. Jon Grinspan, "'Sorrowfully Amusing': The Popular Comedy of the Civil War," *Journal of the Civil War Era* 1 (September 2011): 313–38 (Lincoln quote on 323).
11. For these examples and more, see Rebecca Solnit's crucial book *A Paradise Built in Hell: The Extraordinary Communities That Arise in Disaster* (New York: Penguin, 2009), 13–70 (quotations on 26 and 15).
12. Quoted in Kevin Rozario's excellent book *The Culture of Calamity: Disaster and the Making of Modern America* (Chicago: University of Chicago Press, 2007), 101.
13. Thornton Wilder, *The Skin of Our Teeth* (1942; repr., New York: Samuel French, 1944), 46.
14. Wilder, 8–9, 21.
15. Wilder, 53, 12, 122.
16. Wilder, 7.
17. Mark Polizzotti, "Introduction: Laughter in the Dark," in Breton, *Anthology of Black Humor*, viii–x.
18. Tom Lehrer, *An Evening Wasted with Tom Lehrer* (Lehrer Records, 1959); Lehrer, *That Was the Year That Was* (Reprise Records, 1965). Lehrer has long been extremely generous in granting permission to anyone who wishes to quote his lyrics (without charge), but he is quite

particular about his line breaks, of which he is justifiably proud. The line breaks used here come directly from him.

19. Edward Abbey, "Down the River with Henry Thoreau," in *The Best of Edward Abbey* (San Francisco: Sierra Club Books, 1988), 272–307 (quotation on 279).
20. Edward Abbey, *Desert Solitaire: A Season in the Wilderness* (1968; repr., New York: Ballantine, 1971), 71–72, 20. Also see Daegan Miller, "On Possibility; or, The Monkey Wrench," in *Future Remains: A Cabinet of Curiosities for the Anthropocene*, ed. Gregg Mitman, Marco Armiero, and Robert S. Emmett (Chicago: University of Chicago Press, 2018), 141–48. The twenty-first-century inheritor of Abbey's mantle is Michael P. Branch; see, for instance, *Raising Wild: Dispatches from a Home in the Wilderness* (Boulder, CO: Shambhala, 2016).
21. Michael Pollan, *Second Nature: A Gardener's Education* (New York: Delta, 1991), 4, 52–53.
22. James Russell, "Climate Change? You're Having a Laugh," *Guardian*, May 7, 2008, www.theguardian.com.
23. See, for instance, Dipesh Chakrabarty, "The Climate of History: Four Theses," *Critical Inquiry* 35 (Winter 2009): 197–222; and Chakrabarty, "Climate and Capital: On Conjoined Histories," *Critical Inquiry* 41 (Autumn 2014): 1–23.
24. RSA, "Marcus Brigstocke on Climate Change," YouTube, January 30 2015, www.youtube.com/watch?v=KDUcQY3jO4M&t=307s.
25. Jenny Price, "Green Me Up, JJ," *Native Intelligence* (blog), *LAObserved*, January 23, 2010, www.laobserved.com.
26. The Yes Men, "Hijinks/RefuGreenErgy," September 1, 2017, www.theyesmen.org.
27. The Yes Men, "Hijinks/Shell's Last Iceberg Snow Cones," June 1, 2015, www.theyesmen.org.
28. The Yes Men, "Reed Divests / Reed College Commencement Speech 2014," May 19, 2014, www.theyesmen.org.
29. Yannis Nikolakopoulos, "Australians for Coal. What Is Your Investment Dollar Doing?," YouTube, March 1, 2014, www.youtube.com/watch?v=tqXzAUaTUSc.

30. This is a classic case of environmental injustice. On this particular situation, see, for instance, "Wangan and Jagalingou Traditional Owners to Appeal after Losing Critical Adani Case," National Indigenous Television, Special Broadcasting Service (Australia), August 17, 2018, www.sbs.com.au. More broadly, see Aaron Sachs, *Eco-Justice: Linking Human Rights and the Environment*, Worldwatch Paper 127 (Washington, DC: Worldwatch Institute, 1995); Daniel Faber, ed., *The Struggle for Ecological Democracy: Environmental Justice Movements in the United States* (New York: Guilford, 1998); Luke W. Cole and Sheila R. Foster, *From the Ground Up: Environmental Racism and the Rise of the Environmental Justice Movement* (New York: New York University Press, 2001); Julian Agyeman, *Sustainable Communities and the Challenge of Environmental Justice* (New York: New York University Press, 2005); Rob Nixon, *Slow Violence and the Environmentalism of the Poor* (Cambridge, MA: Harvard University Press, 2011); Aaron Sachs, "Looking Backward (Not Forward) to Environmental Justice," in *State of the World 2014: Governing for Sustainability*, ed. Tom Prugh and Michael Renner (Washington, DC: Island, 2014), 105–14; Tracey Skillington, *Climate Justice and Human Rights* (New York: Palgrave Macmillan, 2017); David Naguib Pellow, *What Is Critical Environmental Justice?* (Cambridge, UK: Polity, 2018); Julie Sze, *Environmental Justice in a Moment of Danger* (Berkeley: University of California Press, 2020); and Dina Gilio-Whitaker, *As Long as Grass Grows: The Indigenous Fight for Environmental Justice, from Colonization to Standing Rock* (Boston: Beacon, 2019).
31. See the Inside the Greenhouse home page: https://insidethegreenhouse.org.
32. Inside the Greenhouse, "Stand Up for Climate Change on April 22," April 22, 2020, https://insidethegreenhouse.org.
33. Inside the Greenhouse.
34. Funny or Die and Defend Our Future, "Old People Don't Care about Climate Change," April 21, 2016, www.funnyordie.com.
35. See Shiloh R. Krupar, *Hot Spotter's Report: Military Fables of Toxic Waste* (Minneapolis: University of Minnesota Press, 2013), 232–45; and

for more on Nuclia Waste, see her Facebook page. The person behind the persona is David Westman.

36. Climate Stew, "Episode 16: A Queer Response to Climate Change," January 5, 2015, https://climatestew.com.
37. Quoted in Nicole Seymour, *Bad Environmentalism: Irony and Irreverence in the Ecological Age* (Minneapolis: University of Minnesota Press, 2018), 140.
38. See the *Climate Stew* home page for Toscano's clip: https://climatestew.com.
39. *New Yorker*, February 18–25, 2019, 34. Also note the *New Yorker* article of January 28, 2019, by Charles Bethea, about the climate activist Jon Leland: "When Will My Apartment Be Underwater?" Leland printed eight thousand biodegradable stickers saying, "This Place Will Be Water," which he put up all over New York City. A recent book notable for its honesty about the burden of growing up with climate change is Daniel Sherrell, *Warmth: Coming of Age at the End of Our World* (New York: Penguin, 2021).
40. *Colbert Report*, Comedy Central, May 14, 2014, www.cc.com.
41. Henry Feldman, "Plays," Henry Feldman's website, accessed June 30, 2022, www.henryfeldman.me. Also see Alison Rooney, "Sea-Level Rise: The Comedy; Cold Spring Playwright Takes Madcap Approach," *Highlands Current*, July 20, 2019, https://highlandscurrent.org.

EPILOGUE

1. I've been compiling this list for the past few years, as material for my climate-change comedy hours and stand-up routines. But others have been thinking along similar lines; for instance, Ginny Hogan, "Idioms Updated for Climate Change," *New Yorker*, January 21, 2019, 25. One of Hogan's examples: "Ugh, she's giving me the tepid shoulder again."
2. See Brian Fagan, *The Little Ice Age: How Climate Made History, 1300–1850* (New York: Basic Books, 2001); Philipp Blom, *Nature's Mutiny: How the Little Ice Age of the Long Seventeenth Century Transformed the West and Shaped the Present* (New York: Liveright,

2019); and "Meaning and Origin of the Phrase 'Cold Comfort,'" Word Histories, December 26, 2016, https://wordhistories.net.

3. N. M. Tchebakova, Elena I. Parfenova, Galina I. Lysanova, and Amber J. Soja, "Agroclimatic Potential across Central Siberia in an Altered Twenty-First Century," *Environmental Research Letters* 6 (November 2011), http://iopscience.iop.org.
4. There is actually some science to support this claim. There is also some science to support the opposite claim. Apparently, it depends on how you define "cold-related death" and "heat-related death." Details! See, for instance, Veronika Huber, "Will Climate Change Bring Benefits from Reduced Cold-Related Mortality? Insights from the Latest Epidemiological Research," *RealClimate*, June 11, 2018, www.realclimate.org.
5. Yes, I know that we're supposed to hate Woody Allen now, but he's just too much a part of the culture I grew up in—I can't avoid him.
6. Amitav Ghosh makes some of these points in his book *The Great Derangement: Climate Change and the Unthinkable* (Chicago: University of Chicago Press, 2016), 147–49. Also note that Beth Osnes, at the University of Colorado, has devoted a great deal of scholarly and artistic energy to connecting comedy and resilience in the context of climate change; see her recent book *Performance for Resilience: Engaging Youth on Energy and Climate through Music, Movement, and Theatre* (New York: Palgrave Macmillan, 2017). She starts the book with a nod to Tom Lehrer.
7. I can't find an original instance for this quotation, but many online tributes to Gregory cite it as one of his early taglines.
8. Aristophanes, *Frogs*, in *The Complete Plays of Aristophanes* (New York: Bantam Books, 1971), 372–73.
9. Lateshia Beachum, "'We Took Hot Girl Summer Too Far': The Best Signs from the Global Climate Protests," *Washington Post*, September 20, 2019, www.washingtonpost.com.
10. Between 2015 and 2017, several articles about pre-traumatic stress appeared in publications like *In These Times*, *Psychology Today*, *U.S. News*, and *The Huffington Post*. See, for example, Martin de

Bourmont and Dayton Martindale, "Is Climate Change Causing Pre-Traumatic Stress Disorder in Millennials?," *In These Times*, August 10, 2015, http://inthesetimes.com. Also note the theorization of "Pretraumatic Stress Syndrome" in E. Ann Kaplan, *Climate Trauma: Foreseeing the Future in Dystopian Film and Fiction* (New Brunswick, NJ: Rutgers University Press, 2016). Also see "Report: More U.S. Soldiers Suffering from Pre-Traumatic Stress Disorder," *The Onion*, November 15, 2006, www.theonion.com.

11. Chris Rasmussen, "'This Thing Has Ceased to Be a Joke': The Veterans of Future Wars and the Meanings of Political Satire in the 1930s," *Journal of American History* 103 (June 2016): 84–106.
12. dubeyeo8, "George Carlin Shell Shock," YouTube, August 4, 2012, www.youtube.com/watch?v=hSp8IyaKCso.
13. See Mary L. Dudziak, *War Time: An Idea, Its History, Its Consequences* (New York: Oxford University Press, 2012).
14. Rasmussen, "This Thing Has Ceased to Be a Joke," 85, 104.

INDEX

Page numbers in italics indicate illustrations

ABOUT THE AUTHOR

Aaron Sachs is Professor of History and American Studies at Cornell University. He is the author of *The Humboldt Current: Nineteenth-Century Exploration and the Roots of American Environmentalism*, *Arcadian America: The Death and Life of an Environmental Tradition*, and *Up from the Depths: Herman Melville, Lewis Mumford, and Rediscovery in Dark Times*. With John Demos, he co-edited *Artful History: A Practical Anthology*.